Nanotechnology Science and Technology

FROM GOLD NANO-PARTICLES THROUGH NANO-WIRE TO GOLD NANO-LAYERS

NANOTECHNOLOGY SCIENCE AND TECHNOLOGY

Safe Nanotechnology
Arthur J. Cornwelle
2009. 978-1-60692-662-8

National Nanotechnology Initiative: Assessment and Recommendations
Jerrod W. Kleike (Editor)
2009. 978-1-60692-727-4

Nanotechnology Research Collection - 2009/2010. DVD edition
James N. Ling (Editor)
2009. 978-1-60741-293-9

Strategic Plan for NIOSH Nanotechnology Research and Guidance
Martin W. Lang
2009. 978-1-60692-678-9

Safe Nanotechnology in the Workplace
Nathan I. Bialor (Editor)
2009. 978-1-60692-679-6

Nanotechnology in the USA: Developments, Policies and Issues
Carl H. Jennings (Editor)
2009. 978-1-60692-800-4

Nanotechnology: Environmental Health and Safety Aspects
Phillip S. Terrazas (Editor)
2009. 978-1-60692-808-0

New Nanotechnology Developments
Armando Barrañón (Editor)
2009. 978-1-60741-028-7

Bio-Inspired Nanomaterials and Nanotechnology
Yong Zhou (Editor)
2009. 978-1-60876-105-0

Nanofibers: Fabrication, Performance, and Applications
W. N. Chang (Editors)
2009. 978-1-61668-288-0

Nanotechnology: Nanofabrication, Patterning and Self Assembly
Charles J. Dixon and Ollin W. Curtines (Editors)
2010. 978-1-60692-162-3

Gold Nanoparticles: Properties, Characterization and Fabrication
P. E. Chow (Editor)
2010. 978-1-61668-009-1

**Micro Electro Mechanical Systems (MEMS):
Technology, Fabrication Processes and Applications**
Britt Ekwall and Mikkel Cronquist (Editors)
2010. 978-1-60876-474-7

Nanomaterials: Properties, Preparation and Processes
Vinicius Cabral and Renan Silva (Editors)
2010. 978-1-60876-627-7

**Nanopowders and Nanocoatings:
Production, Properties and Applications**
V. F. Cotler (Editor)
2010. 978-1-60741-940-2

**Nanomaterials Yearbook - 2009 . From Nanostructures,
Nanomaterials and Nanotechnologies to Nanoindustry**
Gennady E. Zaikov and Vladimir I. Kodolov (Editors)
2010. 978-1-60876-451-8

**Nanoparticles: Properties, Classification,
Characterization, and Fabrication**
Aiden E. Kestell and Gabriel T. DeLorey (Editors)
2010. 978-1-61668-344-3

Nanoporous Materials: Types, Properties and Uses
Samuel B. Jenkins (Editor)
2010. 978-1-61668-182-1

Mechanical and Dynamical Principles of Protein Nanomotors: The Key to Nano-Engineering Applications
A. R. Khataee and H. R. Khataee
2010. 978-1-60876-734-2

Electrospun Nanofibers and Nanotubes Research Advances
A. K. Haghi (Editor)
2010. 978-1-60876-762-5

TiO2 Nanocrystals: Synthesis and Enhanced Functionality
Ji-Guang Li , Xiaodong Li, Xudong Sun
2010. 978-1-60876-838-7

Nanomaterial Research Strategy
Earl B. Purcell (Editor)
2010. 978-1-60876-845-5

Magnetic Pulsed Compaction of Nanosized Powders
G.Sh. Boltachev, K.A. Nagayev, S.N. Paranin,
A.V. Spirin and N.B. Volkov
2010. 978-1-60876-856-1

Nanostructured Conducting Polymers and their Nanocomposites: Classification, Properties, Fabrication and Applications
Ufana Riaz and S.M. Ashraf
2010. 978-1-60876-943-8

Phage Display as a Tool for Synthetic Biology
Santina Carnazza and Salvatore Guglielmino
2010. 978-1-60876-987-2

Bioencapsulation in Silica-Based Nanoporous Sol-Gel Glasses
Bouzid Menaa, Farid Menaa, Carla Aiolfi-Guimarães
and Olga Sharts
2010. 978-1-60876-989-6

ZnO Nanostructures Deposited by Laser Ablation
M. Martino, D. Valerini, A.P. Caricato,
A. Cretí, M. Lomascolo, R. Rella
2010. 978-1-61668-034-3

Development and Application of Nanofiber Materials
Shou-Cang Shen, Wai-Kiong Ng, Pui-Shan Chow
and Reginald B.H. Tan
2010. 978-1-61668-931-5

Polymers as Natural Composites
Albrecht Dresdner and Hans Gärtner (Editors)
2010. 978-1-61668-168-5

Synthesis and Engineering of Nanostructures by Energetic Ions
Devesh Kumar Avasthi and Jean Claude Pivin (Editors)
2010. 2978-1-61668-209-5

From Gold Nano-Particles Through Nano-Wire to Gold Nano-Layers
V. Švorčík, Z. Kolská, P. Slepička
and V. Hnatowicz
2010. 978-1-61668-316-0

Gold Nanoparticles: Properties, Characterization and Fabrication
P. E. Chow (Editor)
2010. 978-1-61668-391-7

Nanoporous Materials: Types, Properties and Uses
Samuel B. Jenkins (Editor)
2010. 978-1-61668-650-5

Phase Mixture Models for the Properties of Nanoceramics
Willi Pabst and Eva Gregorova
2010. 978-1-61668-673-4

Applications of Electrospun Nanofiber Membranes for Bioseparations
Todd J. Menkhaus, Lifeng Zhang and Hao Fong
2010. 978-1-60876-782-3

Development and Application of Nanofiber Materials
Shou-Cang Shen, Wai-Kiong Ng, Pui-Shan Chow
and Reginald B.H. Tan
2010. 978-1-61668-829-5

Polymers as Natural Composites
Albrecht Dresdner and Hans Gärtner (Editors)
2010. 978-1-61668-886-8

Phase Mixture Models for the Properties of Nanoceramics
Willi Pabst and Eva Gregorova
2010. 978-1-61668-898-1

Nanotechnology Science and Technology

FROM GOLD NANO-PARTICLES THROUGH NANO-WIRE TO GOLD NANO-LAYERS

V. ŠVORČÍK
Z. KOLSKÁ
P. SLEPIČKA
AND
V. HNATOWICZ

Nova Science Publishers, Inc.
New York

NOTICE TO THE READER
The Publisher has taken reasonable care in the preparation of this book, but makes no expressed or implied warranty of any kind and assumes no responsibility for any errors or omissions. No liability is assumed for incidental or consequential damages in connection with or arising out of information contained in this book. The Publisher shall not be liable for any special, consequential, or exemplary damages resulting, in whole or in part, from the readers' use of, or reliance upon, this material.
Independent verification should be sought for any data, advice or recommendations contained in this book. In addition, no responsibility is assumed by the publisher for any injury and/or damage to persons or property arising from any methods, products, instructions, ideas or otherwise contained in this publication.
This publication is designed to provide accurate and authoritative information with regard to the subject matter covered herein. It is sold with the clear understanding that the Publisher is not engaged in rendering legal or any other professional services. If legal or any other expert assistance is required, the services of a competent person should be sought. FROM A DECLARATION OF PARTICIPANTS JOINTLY ADOPTED BY A COMMITTEE OF THE AMERICAN BAR ASSOCIATION AND A COMMITTEE OF PUBLISHERS.

LIBRARY OF CONGRESS CATALOGING-IN-PUBLICATION DATA

Available upon Request
ISBN: 978-1-61668-316-0

Published by Nova Science Publishers, Inc. ✛ New York

CONTENTS

PREFACE

In the course of the 20th century the theory of size-effects in metal thin layers has been explained nebo described by numerous scientists, and various approaches to the problem have been proposed. Behaviour of metal 1D, 2D and 3D particles of "exiguous" dimensions is affected mainly by the surface size and quantum size effects. These phenomena have become subject of recently developed scientific discipline known as Size-Dependent Chemistry. The surface- and size-effects can be attributed to a high nanoparticle surface-to-bulk ratio. Hand in hand with the reduction of nanoparticle dimension, surface atoms' proportion increases dramatically, thus commonly known physical properties of the bulk materials change e.g., density and melting point of Au nanoparticle decreases. Quantum effects are observed in the study of electricial properties of Au nanolayers.

In this contribution, present knowledge on the properties of both Au nanoparticles and Au nanolayers deposited on various substrates is summarized and some recent experimental results obtained on synthetic polymers are presented. Gold particles and layers were evaporated or sputtered on pristine polymer or polymer activated by plasma discharge or by irradiation with laser light. The chemical bonding of Au nanoparticles on the activated polymer surface was demonstrated. Grafting of the polymer surface with thiol group (-SH) leads to formation of an interlayer which makes adhesion of Au nanoparticles easier. It was also shown that Au nanoparticles deposited on the polymer surface increase adhesion and proliferation of living cells dramatically, this fading being important for tissue engineering.

Thickness and density of the gold layers were determined using Atomic Absorption Spectroscopy (AAS), Focused Ion Beam (FIB), Atomic Force Microscopy (AFM) and gravimetry. Surface roughness and morphology of deposited gold nanoparticles and nanolayers were examined by AFM, Scanning electron microscopy (SEM) and Transmission electron microscopy (TEM). Electrical properties of polymer–gold structures such as temperature dependence of resistance, C-A characteristic, charge transport mechanism, concentration and mobility of free charge cariers were studied by conventional methods and optical properties such as optical gap and center oscillator energy were deduced. Gold-crystal structure was determined from X-ray Diffraction (XRD) measurements. Biological properties (e.g. cells adhesion and proliferation) were studied on cell cultures *in vitro* experiments.

AUTHOR AFFILIATIONS

V. Švorčík
Department of Solid State Engineering, Institute of Chemical Technology, 166 28 Prague, Czech Republic, vaclav.svorcik@vscht.cz

Z. Kolská
Department of Chemistry, J.E. Purkyne University, 400 96 Usti nad Labem, Czech Republic, zdenka.kolska@ujep.cz

P. Slepička
Department of Solid State Engineering, Institute of Chemical Technology, 166 28 Prague, Czech Republic

J. Siegel
Department of Solid State Engineering, Institute of Chemical Technology, 166 28 Prague, Czech Republic

V. Hnatowicz
Institue of Nuclear Physics, Academy of Sciences of the Czech Republic, 25068 Rez, Czech Republic, hnatowicz@ujf.cas.cz

Chapter 1

INTRODUCTION

Vacuum-deposited metal clusters and thin films play an important role in various fields of technology [1]. Metallized plastics are nowadays widely used in products ranging from e.g., car reflectors, compact discs, electric shielding to foils for food packing and biosensors [2-4]. The polymer/metal structures are a base for the construction of diodes with negative differential resistance [5], light emitting diodes in optoelectronics [6], organic transistors [7,8] or MEMS devices [9]. Metallized polymer foils are the initial materials for the preparation of active food packing for microwave warming [10]. The polymer-metal or polymer-carbon composites, prepared by penetration of metal or carbon particles at elevated temperature [11,12], are intensively studied as a parent material for the construction of moisture sensors and optical switches [13]. Wide range of applications of metallized polymers has stimulated research of metal-polymer interaction. The mechanical, optical and electric properties of the metal-polymer composites are affected by the technique of the layer deposition, metal reactivity and the structure of metal-polymer interface which in turn is influenced by the degree of metal diffusion and metal-polymer intermixing [3]. The polymer metallization is accomplished using sputtering, evaporation and Chemical Vapour Deposition (CVD) techniques [5,12-14]. The formation of the metallic ad-layers is governed by complex nucleation processes, which are still not completely understood [15]. Several techniques have been developed for measuring of the thickness, surface continuity and morphology of the metal layers deposited on polymers [16]. While for certain applications the continuous metallic coverage are required (e.g. as contacts for electrical measurements or MEMS [8,15]), in other cases, discontinuous metallic films are needed (e.g. for diodes with negative effective resistance [5] or food packaging for microwave heating [10]).

High-dose implantation of metal ions into polymers leads to the local excess of the dopant, the system then relaxes into metal precipitates–colloids, granules or nano-particles. The size of particles, concentration and the depth of implanted region can be controlled to some extent by the regimes of the ion implantation. In this way the different kinds of metal nano-particles may be formed in the near-surface layer of the irradiated dielectrics. Studies of nano-size particles and clusters are very rapidly growing area of research. The physical properties of matter on nanoscale are essentially different from the bulk ones reflecting inherent property of substance to be quantized. Unique phenomena observed on nanoscale allow tailoring new nano-size and nanostructured materials with novel characteristics. Some new physics has already realised in modern technology applications, especially in those of them that utilise magnetic and electronic properties of the substances for magnetic recording and information storage, magnetosensor electronics, magneto-optical devices, ferrofluids, etc. [17].

Adhesion of the metal layer is of great importance in most of applications. However, the adhesion between metals and polymers is normally very poor, causing difficulties when exploring applications of the metallized polymers. Metals and polymers are rather dissimilar materials, e.g., the cohesive energy of metals is typically two orders of magnitude higher than that of polymers. This is one reason for very weak interaction between metals and polymers. Another reason is that the polymer surfaces are usually nonpolar, or have very low wettability [18].

Surface modification techniques, which can transform inexpensive polymers into highly valuable finished products, have become an important part of the plastics and many other industries. Many advances have been made in surface treatment techniques to alter the chemical and physical properties of polymer surfaces without affecting polymer bulk properties. It has been observed earlier that exposing the polymer substrate to plasma discharge or to ionising radiation [19,20] can enhance adhesion of metal layers. The adhesion could favourably be affected also by changes in polymer surface morphology resulting from e.g. plasma treatment.

A number of parameters are involved in the layer growth process, such as electronic structures and surface free energies of the two materials involved, polymer surface morphology, polymer molecular chain mobility and experimental deposition conditions, determining the size, shape, and atomic structure of the metal clusters and the respective temporal correlations [1]. In turn, film morphology influences the characteristics of the interface and, consequently, the response of the whole system. In the initial stage of film growth, metal atoms can diffuse easily into the polymer and form embedded clusters, presupposing that the deposition temperature is close to or above the glass transition temperature of the polymer [21,22]. Interdiffusion at the interface is in favor of a high adhesion strength [23]. On the other hand, penetrating metal atoms may alter the electronic properties at the interface and therefore make it difficult to control the properties of an electronic device based on polymer-metal composites [24].

The ionising radiation or plasma discharge create new structures on the polymer surface (e.g. radicals, conjugated double bonds, oxygen containing groups) which may facilitate metal deposition and bonding. Ion irradiation of metal-polymer composites may stimulate diffusion and atomic mixing on metal-polymer interface. The issue is however still an open question, which demands further investigation. The structure and morphology of the deposited layers depend on the deposition technique and on the affinity of the polymer substrate to particular metal. Interfacial characteristics of the metal-polymer composites are of importance from the point of view of the spontaneous deterioration of the metal-polymer composite systems. As the process of metal diffusion into polymer may affect the ageing of the system, the diffusion of metals in polymers is an important research topic.

Despite of the fact that the metal diffusion has been studied in various metal-polymer systems the diffusion mechanisms are far from being understood. Nevertheless, strong correlation has been observed between metal reactivity and its diffusivity which, however, is by many orders of magnitude lower than that of gas molecules of comparable size [25]. Obviously, metal diffusion into polymers is strongly impeded by metal proclivity to aggregate. This is perhaps the reason why no significant diffusion has been observed for the systems with continuous metalloid coverage [25]. Moreover, it is not clear what the mechanism of the metal diffusion in polymers is. Several models including atomic mode diffusion [26] or 'diffusion' of the nano-particles or gas molecules via free volume [27] are currently discussed.

The interaction of metals with organic functional groups in thin films is important in a number of areas including organometallic chemistry, polymer chemistry and more recently to the growing field of molecular electronics. During the past few years, a number of photoemission studies [28–30] have been performed to investigate the chemical bonding in polymer surfaces and metal–polymer interfaces. For the study of interfacial chemical reaction in polymer-metal system X-ray photoelectron spectroscopy (XPS) can advantageously be used. However, the details of chemical bonding at the surface and the interface in polymer-metal system are not adequately known yet.

The two-step procedure comprising the metal deposition onto polymer surface and subsequent thermal annealing of the metal-polymer sandwiched systems is one of the possible techniques for the introduction of metals (or metallic nano-particles) into polymers. Alternative approaches used for the fabrication of the polymer-based nano-composites, e.g., the blending of polymers with commercially available metallic nano-particles, ion implantation, seem to be also promising [31], however they are still in the early stages of exploration. Polymer-metal nano-composites have emerged in the recent years as an important issue due to their practical and fundamental significance [32,33]. The metallic nano-particles embedded in the insulating host exhibit characteristic non-linear optical, electric and magnetic properties, with possible applications in luminescent and non-linear optical devices [34,35].

Polymer-metal composites potentially offer a remarkable combination of properties, i.e. the effectiveness of polymer as coatings and frame-structural material, combined with versatile optical, electric, magnetic and chemical properties bestowed by the nano-particles. Properties of such composites will depend on the size and organization of nano-particles inside the polymer matrix. From a more fundamental point of view, the metal nano-particles embedded in the insulating material can also be considered as quantum dots or artificial atoms exhibiting peculiar electrical and magnetic properties [35]. Electrical properties of metal-insulator-metal composites (MIM) with inorganic insulators have been intensively studied and their properties are well known [36]. On the other hand, little is known on the electrical properties of MIM structures with polymeric central layer and some of effects (e.g. the appearance of negative differential resistance) are waiting for experimental verification and theoretical explanation.

Artificial organic and inorganic materials, such as polymers, metals and ceramics, have increasingly been applied in medicine and in various biotechnologies, such as bioimaging, biosensing, drug delivery, cell cultivation and also in constructing replacements for irreversibly damaged tissues and organs. In the case of replacements, the material should not be merely passively tolerated by the cells but should as far as possible promote specific cell responses in a controllable manner. For example, materials serving as carriers of cells for tissue engineering and tissue regeneration should behave by analogy with extracellular matrix (ECM). That is, they should regulate the extent and strength of cell adhesion, the subsequent proliferation activity of cells, switch between a proliferation and differentiation program in cells, cell maturation and performance of specific functions typical for the given cell type.

Interaction of cells with a polymer substrate is a complex process. Several chemical and physical factors affect the bio-recognition process of the specific binding between the cell surface receptors and their corresponding ligands. Many studies have focused on improving the affinity between cells and a material surface, the results of which are summarized in reviews [37,38].

An important feature that renders an artificial material bioactivity is its nanostructure. The nanostructure of the material surface, i.e., the presence of irregularities smaller than 100 nm, is considered to mimic the topography of the cell membrane as well as the folding of physiological ECM molecules [39,40]. It is known that cell adhesion to artificial materials is mediated by ECM molecules spontaneously adsorbed to the material surface. In a cell culture system, these molecules are provided by the serum supplement of the culture media; under *in vivo* conditions, from the body fluids, and in both systems, ECM molecules are synthesized by the cells themselves. On nanostructured surfaces, it is believed that ECM molecules are adsorbed in an appropriate amount and spectrum, and particularly in an appropriate geometrical conformation needed for the accessibility of specific amino acid sequences in these molecules (e.g., Arg-Gly-Asp, RGD) by cell adhesion receptors, i.e. integrin and non-integrin adhesion molecules present on the cell membrane [41,42].

Alternative methods for creating or further manipulating the nanoscale surface roughness of polymers include physical methods for the material surface modification, such as exposing to plasma discharge, or irradiation with ultraviolet (UV)-light or ions [43,44]. A common feature of all these approaches is degradation of the polymer macromolecules and an increase in the nanoscale surface roughness [43].

Another important feature of the modification is the formation of free radicals and their subsequent reaction with oxygen in the ambient atmosphere. The newly formed oxygen-containing chemical functional groups render the material surface more wettable, which supports the adsorption of ECM molecules in a flexible and reorganisable form. This form enhances the accessibility of oligopeptidic ligands on these molecules for cell adhesion receptors [41,44,45]. Thus, the appropriate wettability of the polymer surface acts synergetically with the surface nanoscale roughness on improving of the cell adhesion. Another interesting property of irradiated polymers supporting their colonization with cells is the formation of conjugated double bonds between carbon atoms and increased electrical conductivity of the material, which has also been shown to have positive effects on the adhesion, growth and maturation of various cell types, including vascular smooth muscle and endothelial cells [46].

In addition, free radicals formed on irradiated polymers can be utilized for attaching non-toxic and biocompatible nanoparticles on the material surface in order to create artificial bioinspired nanostructured surfaces for tissue engineering. Among metallic materials, gold, including its nano-sized form, has shown no cytotoxicity *in vitro* and *in vivo*. For example, gold nanoparticles, although internalized into cultured human hepatocellular and pancreatic cancer cells, have had no impact on the proliferation and mitochondrial activity of these cells [47]. Similarly, mouse embryonic stem cells maintained their ability to proliferate for several passages while grown in the presence of gold nanoparticles. In addition, one month after intravitreal injection of nanogold, rabbits showed no signs of retinal or optic nerve damage [48].

The non-toxicity of gold is related to its well-known stability, non-reactivity and bioinertness. However, gold can easily react with thiol (-SH) derivates giving Au-S bond formation. Gold nanoparticles can be functionalized with polyfunctional thiols containing vinyl, hydroxyl, carboxyl and amine groups, which are known to react with radicals in solution [49], and thus they can be expected to react also with radicals on the polymer surface radicalized by irradiation with plasma, ultraviolet light from an excimer lamp or ion irradiation [44-46]. Gold nanoparticles of various sizes will also enable one to define the optimum size and shape of material surface irregularities for cell adhesion and growth, because potential differences in cell responses to the material topography within the scale from a few (e.g. 3) to 100 nm have not yet been systematically studied.

Thus, newly constructed surfaces will combine advantageously several features, which will synergetically enhance the attractiveness of the material surface for cell colonization: nanostructure, appropriate chemical composition, wettability, electrical charge and conductivity.

As mentioned above, the free vinyl and oxygen- and nitrogen-containing groups on the material surface can be utilized for further functionalization by the molecules (e. g. oligopeptidic ligands) for cell adhesion receptors. These ligands can contain RGD sequences recognized by a variety of cell types or amino acid sequences bound preferentially by a certain cell type, such as e.g. Lys-Gln-Ala-Gly-Asp-Val or Val-Ala-Pro-Gly recognized by vascular smooth muscle cells (VSMC) or Arg-Glu-Asp-Val specific for endothelial cells (for a review, see [6,15]). Moreover, the concentration and spatial distribution of the adhesion ligands can regulate the extent and strength of cell adhesion, cell shape and the subsequent behaviour of cells, including their proliferation and differentiation activity [42,50].

The described material surface modifications, especially those with gold nanoparticles, can also be developed on hard materials in order to be applicable in hard tissue surgery. Conventional approaches involve the creation of nanogold-modified surfaces in hard substrates by dipping or spin-coating techniques on hard substrates [51]. A promising alternative is depositing of synthetic polymers on metallic material currently used for bone implantation and joint replacements, and modifying these polymers with gold nanoparticles and functionalization of biomolecules, as described above.

Chapter 2

PROPERTIES OF AU NANO-PARTICLES

Due to the modern experimental techniques permitting analysis of nano-sized metal particles it has been shown that nanomaterial exhibit different properties in comparison with bulk material [52-57]. Samples 1-100 nm in size represents intermediates between atoms and bulk. Metallic nano-particles have a wide range of applications in many branches as mentioned in Introduction and thorough study of their properties is of great importance from fundamental and practical point of view. Peculiar properties of nano-structured materials can be demonstrated on gold nano-particles or nano-layers. Gold layers, about 10 nm thick and Au nano-particles appear red and smaller ones can even fluorescence [53,55]. Various physico-chemical properties of gold (optical, electrical, mechanical or magnetic ones) strongly depend on the sample size. The electrical properties vary from the metallic behaviour (for bulk) into insulator (for nanoparticle). In contrast to the bulk, gold nano-particles exhibit magnetic properties, the chemical activity increases (they may serve as catalyst [53]), distinct melting temperature [53,55,58,59], latent heat of melting [59,60], interfacial tension [61-63], density and cohesive energy [64].

Thermodynamic description of nanosystems stems from the basic termodynamic relations known for macrosystems. It can be applied to individual particles, clusters of all shapes, wires and nanolayers. As a result of the particle size reduction a relative number of atoms on the particle surface increases significantly. These atoms exhibit different properties (thermal vibration, binding energy, ect.) in comparison with the properties of bulk ones. With increasing fraction of the surface atoms, the importance of surface related quantities such as a surface work or a surface tension, which play a role in formation of nano-particles from individual atoms, increases dramatically.

The smaller nano-system the more pronounced effect of the above mentioned phenomena [54]. Phase transitions are connected with collective behaviour of many particle systems. If the particle number is low, the phase transition is not sharp. Clusters containing very small number of atoms or molecules do not behave as "material" but more likely as single molecules. There are numerous concepts of thermodynamics which can break down, in particular when the system of interest consists of a single isolated cluster with a small number of atoms [54]. In the case of curved interphases (particles and wires) a pressure difference exists between concave (inner) and convex (outer) side of the interphase, which is inversely proportional to the curvature radius. This dependence can be expressed by the Young-Laplace equation [54]. Due to this the properties are also different in narrow pores. The solubility of salts in pore-confined water, the melting point and even the critical point of a fluid are therefore greatly reduced.

Peculiar properties of nano-particles may result in their behaviour in living organisms. Smaller particles exhibit higher mobility and solubility and they are distributed in the organism more rapidly. This may lead to different behaviour of the particles of the same material depending on their dimension in comparison with "macro" material. The effects may be two kinds (i) negative one - nanoparticles may more easily, in comparison with microparticles, penetrate in the organism parts where their effect can be negative or (ii) positive one - the mobile nanoparticles may be used as carrier of targeted medicaments or they can help to kill some dangerous cells. Thanks to these possibilities the new research fields as Nanotoxicology, Nanomedicine and Nanosafety have intensively been developed lately [65-67]. Recently different behaviour of gold nanoparticles in living organisms, depending on the particle size, has been observed [68]. Nanoshells possess highly favorable optical and chemical properties for biomedical imaging and therapeutic applications. For example, nanoshells coated with gold can kill cancer tumors in mice. The nanoshells can be targeted to bond to cancerous cells by conjugating antibodies or peptides to the nanoshell surface. By irradiating the area of the tumor with an infrared laser, which passes through flesh without heating it, the gold is heated sufficiently to cause death to the cancer cells [69,70].

For many years scientists have tried to explain the properties of nanosystems [52-64]. All concepts consider two basic size-depending effects: one which is related to the fraction of atoms at the surface, and quantum effects which show discontinuous behaviour due to completion of shells in systems with delocalised electrons [53]. Both of these effects are not independent but they affect nanosystems properties at the same time.

2.1. SURFACE SIZE EFFECTS

The "first" size effects are based on a presumption, that atoms at the surface of a large piece of metal are different from the same atoms inside a bulk. Also the surface atom of a large metal sample is different from the surface atom of the nanoparticle of the same metal. The reason is the fact that the surface atom haves fewer direct neighbour atoms in comparison with bulk ones. Because of this lower coordination and unsaturated bonds, surface atoms are less stabilized than bulk atoms.

The surface of a particle scales with the square of its radius r^2, but its volume scales with r^3. The fraction of atoms at the surface is expressed as a ratio of the surface area and the particle volume and it scales as r^{-1}. Particle properties, e.g. melting and other phase transition temperatures, depend on the surface-to-volume ratio. As the particle size increases, they change continuously and approach bulk value at last [54].

The mentioned size effects may be enhanced by the effects of different structural arrangement of the atoms in small systems. Several authors observed different values of lattice parameters of the evaporated or sputtered metal layers depending on the layer thickness [52,63,71-75]. The difference in the lattice parameter results naturally in the differences in the layer density. The atoms inside the particle are more coordinated and therefore more stable than those at the surface. When the portion of the surface atoms becomes higher, many properties related to an arrangement of atoms, e.g. melting temperature (see figure 1), should change too [53,55,76].

Dependence of the density on the thickness of the metal layers was studied several times [52,71-77]. Density of gold, aluminium, silver and copper layers evaporated on glass surface in dependence on the layer thickness was presented in [52]. For gold layers the density changes from 18.8 g cm^{-3} when the layer thickness is about 117 nm to a mean density of 19.3 g cm^{-3} for the layer thickness of 200 nm and higher [52]. These results correspond to ones obtained in our researches for sputtered Au layers [71]. Other authors studied properties of gold layers evaporated on silicon with the thickness between 1-100 nm [72]. They reported the dependence of the gold lattice parameter on the layer thicknesses (see figure 2a). The results are similar to that obtained for sputtered Au layers [71] (see figure 2b). The lattice parameter a = 0.40784 found for evaporated layer [72] is close to $a = 0.40731$ for the sputtered layer [71] for compact nanolayers. From figures 2 it is evident, that the dependence of the lattice parameter on the layer thickness is linear for higher thicknesses but for thinner layers is non-linear, the curve is steeper [71,72]. The same results were reported earlier for system gold/glass [73]. Similar results were also presented for gold evaporated on the silicon deposited by the 10 nm layer of Cu [74,75] and for niobium thin films deposited on Si [77]. Explanation of the size dependent changes in the nano-particle properties may partly be in different arrangement of crystal lattice, in a certain volume disorder [72]. For extremely thin layers only exclusive peaks (111) have sufficient intensity to be detected [71,72,74,75], for thicker films also the peak (222) or others were found [57,71,74] by the XRD measurements. The reason for the initial enhancement of the lattice parameter is probably a net expansion of the gold lattice plane distance due to surface effects [75].

Changes in the crystal lattice in dependence on the crystal size and the size effects on material properties were also studied in [63]. Most important causes of the changes in the crystal lattice are interface tension and the stress field induced by the excess volume in the grain boundaries. These defects will induce a stress field and make the atoms in crystallites move from their normal lattice sites. The excess volumes and due to this lattice parameter and other coherent properties depend on the size cluster but also on the method of preparation on nanocluster. By these phenomena the different properties of gold layers prepared by evaporation and sputtering could be explained. It will be discussed also in Sec. 4.1.

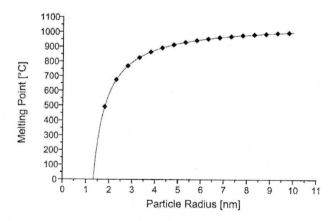

Figure 1. Dependence of the melting point on the Au particle size [76].

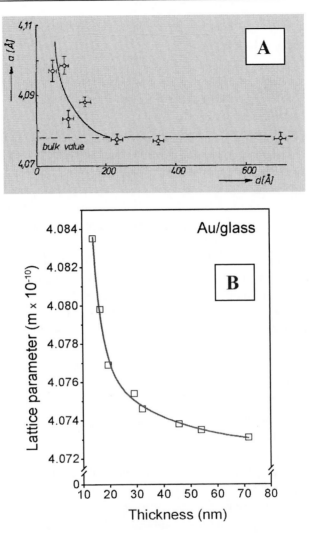

Figure 2. Dependence of the lattice parameters *a* on the layer thicknesses *d* for: A -
evaporated gold layers [72] and B - sputtered gold layers [71].

Very similar results were obtained for different systems studied by other
authors [77]. They discussed the excess free volume, similarly to early work
[63], which is defined as the fractional increase in the molar volume due to
disorder. These excess volume [63] or excess free volume [77], which includes
a fraction of nano-particle volume non-covered by atoms, increase with
decreasing size of the cluster. This effect influences all properties based on the

ratio of atoms mass and cluster volume, which is constant for bulk (19.3 g.cm⁻³) but not for small clusters [71].

2.2. QUANTUM SIZE EFFECTS

The "second" size effects–quantum effects are related to atom orbitals and it is connected with occupation of electron states in systems with delocalized electrons. This effect is important for explanation of differences in electrical and structural properties between nanomaterials and bulk and further for description of the chemical activity, the binding energy between atoms in a particle and between particles, and the crystallographic structure [53,54].

Atoms have their well known atomic orbitals. The core orbitals are confined to a relatively small volume and remain localised (atom-like). Each of atoms contributes with its atomic states to a band so that, although the width of a band increases slightly when more atoms are added, the density of states within a band is basically proportional to the number of atoms of an ensemble with an extended band-like state. The band width amounts typically to a few eV. Thus, the density of states is on the order of atoms number per eV, which is very large for a bulk amount of matter but low for small clusters [53,54].

An important threshold is reached when the gap between the highest occupied and the lowest unoccupied state (called the Kubo gap) equals thermal energy [53,54]. Because of the presence of the Kubo gap in individual nanoparticles, properties such as electrical conductivity and magnetic susceptibility exhibit quantum size effects [59]. When electrons get thermally excited across the Kubo gap, a low temperature insulator becomes a semiconductor and at higher temperatures a metal; and also magnetic properties of small clusters can change dramatically. When we think clusters as molecules, we also should mention bands as delocalised molecular orbitals and the Kubo gap is the HOMO-LUMO gap.

In metal and semiconductor particles the electronic wave functions of conduction electrons are delocalised over the entire particle. Electrons can therefore be described as 'particles in a box', and the densities of states and the energies of the particles depend crucially on the size of the box, which at first leads to a smooth size-dependence. However, when more atoms are added the shells are filled up, and discontinuities occur when a new shell at higher energy starts to be populated. Because of these discontinuities there is no simple scaling. Instead, one finds behaviour akin to that of atoms, with filled shells of extra stability. Therefore, such clusters are often called 'pseudoatoms'. The HOMO–LUMO band gap of semiconductor particles and therefore their absorption and fluorescence wavelengths become size dependent. Ionisation potentials and electron affinities are tuned between the atomic values and the work function of the bulk material by variation of the cluster size. These same properties relate to the availability of electrons for forming bonds or getting involved in redox reactions. Therefore, the catalytic activity and selectivity become functions of size. Quite often, the discontinuous behaviour of quantum size effects is superimposed on a smoothly scaling slope which also reflects the size of a quantised system. It may be difficult to distinguish it from the smoothly scaling surface effect [53].

Next result of the quantum effects is that nanoparticles comprising several hundred atoms of Au, Pd and Pt embedded in a polymer revealed magnetic moments corresponding to several unpaired electron spins per entire particle, even if their bulk materials are not magnetic [78].

Small clusters of transition metal exhibit strong variations as a function of size not only in their physical and electronic properties but also in their chemical behaviour as catalysts [52]. The ability to accept or donate charge plays a key role. One of relevant parameters determining catalytic activity is electron affinity. The electron affinity of gold clusters change by ca 2 eV depending on cluster size. Gold is extremely noble and does not tarnish. This is the reason for its inactivity as a catalyst. However, it has been discovered that small gold particles have excellent catalytic properties [78]. In what follows we discuss the chemical reactivity of metal nanocrystals which is strongly dependent on the size not only because of the large surface area but also a result of the significantly different electronic structure of the small nanocrystals [59]. The electronic structure of a nanocrystal critically depends on its size. For small particles, the electronic energy levels are not continuous as in bulk materials, but discrete, due to the confinement of the electron wavefunction because of the physical dimensions of the particles.

All these discussed properties or effects of properties changes for nanosystems prove significantly during covering of some substrates process. This is important in many branches, where metals deposited on the solid surface, such as glass, polymers, etc. have a wide range of applications. Detailed description of depositing gold on polymer surface, growing the gold layers, etc. is discussed in paper [1]. The growing processes are influenced also by substrate properties and its morphology [1,71]. As can be seen from figure 3, the growing process of the metal layer proceeds in four stages [1]. All of these nanosystem stages have nano-size dimensions and their physical and physico-chemical properties changes with their size as described below.

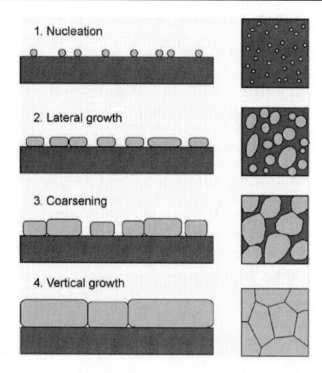

Figure 3. Schematic drawing of gold cluster growth. After the stage of nucleation, cluster growth proceeds mainly in the lateral direction. Coarsening sets in and becomes the dominating process when the surface coverage increases and the clusters get in close contact with each other. With the surface fully covered, adsorption produces only vertical growth and lateral growth is dominated by cluster boundary motion [1].

Chapter 3

PROPERTIES OF AU SPUTTERED
NANO-LAYERS ON SUBSTRATE

3.1. THICKNESS AND DENSITY OF AU NANO-LAYERS

Thickness of the gold layers deposited onto polyethyleneterephthalate (PET) and polytetrafluoroethylene (PTFE), determined from AAS measurement, as a function of the sputtering time is shown in figure 4 [79]. As could be expected, the thickness is an increasing function of the deposition time. For the deposition times below ca 150 s the thicker layers are deposited onto PET (Au/PET), while for longer deposition times the thicker layers are found on PTFE. The difference may be due to different surface morphology of pristine PET and PTFE which is discussed below.

As an alternative method of the thickness determination the FIB method was chosen (see above). The morphology of the FIB cross section is illustrated on SEM image of a Au-PET sample shown in figures 5A,B. Two samples were analyzed with gold layers deposited for 500 and 30 s, with layer thickness 68 and 8 nm respectively. The thickness determined from SEM image for the 500 s sputtering time is in excellent agreement with that determined by AAS method (see figure 4). For the deposition time of 30 s, however, the SEM thickness (8 nm) is twice of that determined indirectly from AAS data (about 4 nm). The difference is obviously due to the fact that the layer deposited for 30 s is not continuous (see figure 5B) and by AAS method only a thickness averaged over large surface area is determined.

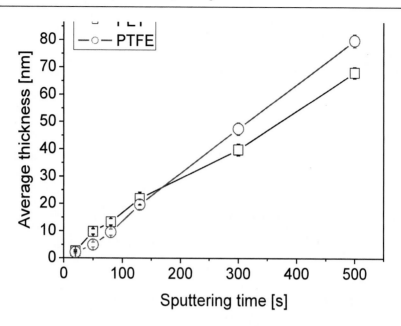

Figure 4. Dependence of the thickness of sputtered Au layer on the Au deposition time for PET and PTFE [79].

Thickness of sputtered layers was measured by AFM, in addition thickness of some layers deposited on glass in initiatory stadium of deposition (sputtering times 50 s) was confirmed by SEM analysis of CP FIB cross-section [71]. Linear dependence between the sputtering time and the layer thickness is evident even in the initiatory stadium of the layer growth, which is in contradiction with results obtained earlier on Au layers deposited on PET [16]. In that case the initiatory stadium of Au layer growth is due to a lower deposition rate.

In figure 6 SEM picture of the cross-section of Au layer at its initiatory stadium of growth is shown. It is obvious, that after 20 s of the deposition flat discrete Au islands (clusters) appear on the substrate surface. After the stage of nucleation growth the growth of Au cluster proceeds mainly in the lateral direction. Coarsening sets in and becomes the dominant process when the surface coverage increases and the clusters get in close contact with each other. After the surface is fully covered, additional adsorption cause only vertical growth and lateral growth is dominated by cluster boundary motion [1].

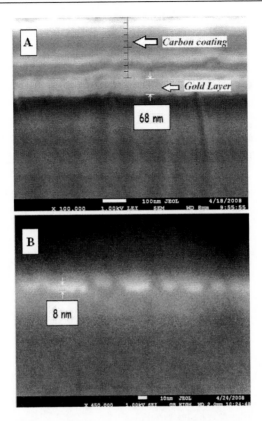

Figure 5. SEM scan of Au layer sputtered on PET for deposition times 500 (A) and 30 s (B). The cut was done by the FIB (focused ion beam) method [79].

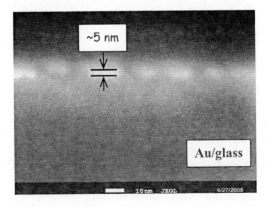

Figure 6. SEM scan of Au layer deposited on glass with deposition time 20 s. The cut was done by the FIB (focused ion beam) method [71].

3.2. ELECTRICAL PROPERTIES OF AU LAYER

Dependence of the sheet resistance (measured by van der Pauw method) on the layer thickness for PET and PTFE can be seen from figure 7 [79]. Because of technical reasons only resistances of the below 106 Ω can be measured in the present experimental set up. With increasing deposition time (and the layer thickness), the sheet resistance of the gold layer decreases rapidly for both polymers. The resistance decrease follows the layer evolution from electrically discontinuous to continuous gold coverage on the polymer surface. In the case of PET (see figure 7) the continuous layer is formed for layer thickness above 10 nm, while for PTFE above 20 nm. The difference may be due to different initial surface morphology of both polymers [79].

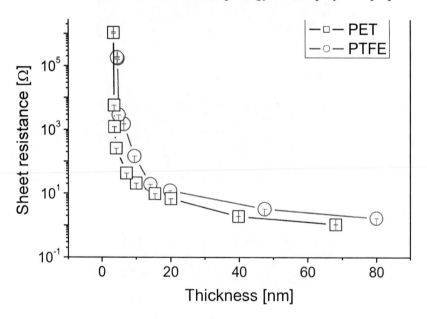

Figure 7. Dependence of the Au layer sheet resistance on layer thickness measured on PET and PTFE with the Van der Pauw technique. Dash-dot lines represent fit of our experimental values [79].

Free carrier volume concentration and their Hall mobility affect significantly electrical conductance of materials. Dependence of free carrier concentration and Hall mobility on the layer thickness for PET and PTFE substrates is shown in figures 8 and 9 [79]. As can be seen from figure 8, the carrier concentration increases dramatically with increasing layer thickness and the layers become conductive (see also figure 7). For both polymers and electrically continuous gold layers the carrier concentration remains constant regardless of the layer thickness. The carrier mobility changes dramatically with increasing layer thickness (figure 9, [79]). The mobility first declines up to the moment when electrically continuous layer is formed. The decline may be due to the fact that in the discontinuous layer the mobility mechanism differs from classical electron conductivity common in metals. Longer deposition time is needed for the formation of continuous Au layer on PTFE than on PET probably due to higher initial roughness of the former one. For thick, electrically continuous gold layers and both polymers the mobility is a slowly increasing function of the layer thickness. In this region the mobility in Au layer on PTFE is significantly lower than that on PET. There is a clear correspondence between mobility (figure 9) and resistance (figure 7) data obtained on PTFE and PET.

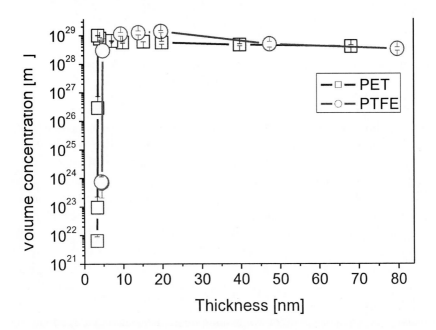

Figure 8. Dependence of volume concentration of free carriers on the thickness of Au layer sputtered on PET and PTFE. The concentration was measured by van der Pauw method [79].

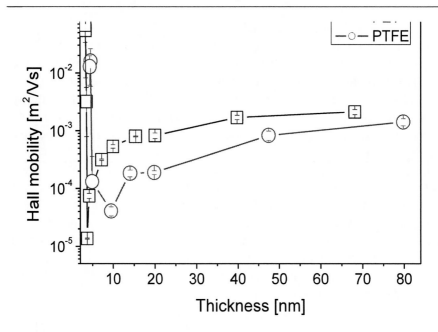

Figure 9. Dependence of free carrier Hall mobility on the thickness of Au layer sputtered on PET and PTFE measured by van der Pauw method [79].

Figure 10 shows the dependence of the sheet resistance of Au layer on the sputtering time. Precedence was given to the dependence on the sputtering time since the accuracy of AFM thickness determination, used in this case, is limited for short sputtering times. It is well known that a rapid decline of sheet resistance of sputtered layer indicates transition from electrical discontinuous to the electrical continuous layer [80]. One can see, that the most pronounced change in the sheet resistance occurs in sputtering time interval 20-50 s, corresponding to 5-10 nm range of the layer thickness. Thus, the layers with the thickness below 5 nm can be considered as discontinuous ones, while the layers with the thickness above 10 nm are definitely continuous. From the measured sheet resistance (figure 10) and effective layer thickness it is possible to calculate the layer resistivity R (Ω cm). One can see that the layer resistivity is about one order of magnitude higher than that reported for metallic bulk gold ($R_{Au} = 2.5 \cdot 10^{-6}$ Ω cm) [81]. The higher resistivity of thin gold layers is due to the size effect in accord with Mattheissen rule [82].

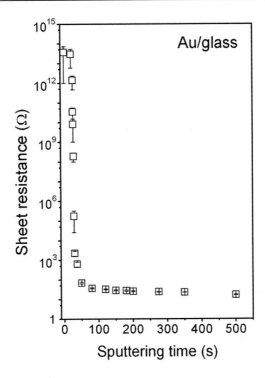

Figure 10. Dependence of the sheet resistance of Au layer sputtered on glass substrate on deposition time [71].

The temperature dependence of the sheet resistance for two particular layer thicknesses is displayed in figure 11. One can see that the temperature dependence of the sheet resistance strongly depends on the layer thickness. For the layer about 89 nm thick the resistance is an increasing function of the sample temperature, the behaviour being expected for metals. For the layer about 6 nm thick the sheet resistance first decreases rapidly with increasing temperature but above the temperature of about 250 K a slight resistance increase is observed. The initial decrease and the final increase of the sheet resistance with increasing temperature are typical for semiconductors and metals respectively. It has been referred elsewhere [52], that a small metal cluster can exhibit both metal and semiconductor characteristics just by varying the temperature. This is due to temperature affected evolution of band gap and density of electron states in the systems containing low number of atoms. From the present experimental data it may be concluded that for the thicknesses above 10 nm the sputtered gold layers exhibit metal conductivity. In the thickness range from 5 to 10 nm the semiconductor and metal conductivity is observed at low and high temperatures respectively. The layers thinner than 5 nm exhibit semiconductive character in the whole investigated temperature scale.

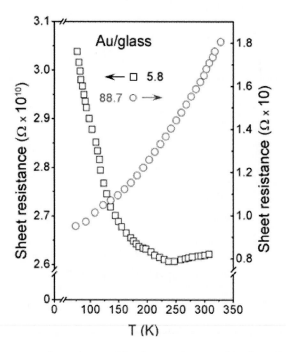

Figure 11. Temperature dependence of the sheet resistance of Au layer deposited on glass substrate for two different layer thicknesses (in nm) as indicated in the figure [71].

Figure 12 displays current-voltage (CV) characteristic of 5.8 nm thick Au layer measured at room temperature (RT) and at the temperature of 90K (LN$_2$). CV curve at RT is strictly linear, so that the Ohm law is valid and the layer exhibits metallic behaviour. CV curve obtained at 90 K grows exponentially, so that it has non-Ohmic character typical for semiconductors. This is in a good accordance with the data of figure 11 and the theory of band gap occurrence in metal nano-structures. While at RT the thermal excitation is big enough for electrons to overcome band gap, at 90 K liquid the band gap can not be overcome. CV dependence measured at RT and 90K on 5.8 nm thick Au layer confirmed former interpretation of the temperature dependence of the sheet resistance, i.e. metallic character of the conductance at RT and semiconductor one at low temperatures.

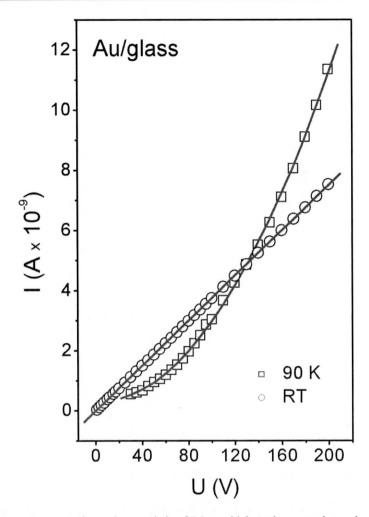

Figure 12. Current-voltage characteristic of 5.8 nm thick Au layer on glass substrate measured at room temperature (RT) and at the temperature of liquid nitrogen (90 K) in cryostat [71].

The temperature dependence of the sheet resistance R_s measured on 5.8 nm thick layer (figure 11) is shown in figure 13 in R_s vs. $T^{1/4}$ representation. One can see that in the temperature range from 90 up to 240 K, the experimental data follow the $R_s \approx \exp[(T_0/T)^{1/4}]$ dependence typical for the variable range hopping mechanism of the electrical conductance, suggested by Mott [83]. Similar behaviour has been reported also for e.g. degraded polymer [84].

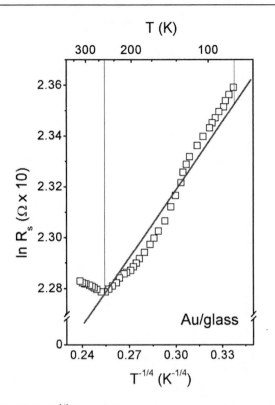

Figure 13. The (ln R_s) vs. $T^{1/4}$ curve (T is sample temperature) for the 5.8 nm Au layer on glass [71].

3.3. OPTICA OPTICAL PROPERTIES OF AU LAYER

Thin Au films exhibit structure-dependent UV-Vis optical spectra [80,85,86]. The localized absorption characteristic of Au films is highly sensitive to the surrounding medium, particle size, surface structure and shape [87]. Transmission spectra from the samples with gold layers of various thicknesses are shown in the figure 14. Only the samples with the gold layer with the thickness below 20 nm, transmitting primary light beam enough, were examined. The spectra exhibit absorption minimum around 500 nm which is slightly red shifted with increasing film thickness. Pronounced absorption at longer wavelength could be attributed to surface plasmon resonance. Discontinuous and inhomogeneous layers, with the thickness ranging from 2.4-9.9 nm and composed of nanometer-sized metal clusters, exhibit absorption in the visible region attributed to the surface plasmon excitation in the metal islands. The surface plasmon peak is shifted from 720 to 590 nm as the nominal layer thickness decreases from 19.5 to 2.4 nm. It is well known, that the optical absorption of island films of gold is a function of island density [88]. The absorption band resulting from bounded plasma resonance in the particles is shifted to longer wavelengths as the island density increases. As the thickness become greater the absorption band is broadened due to wider particles size distribution.

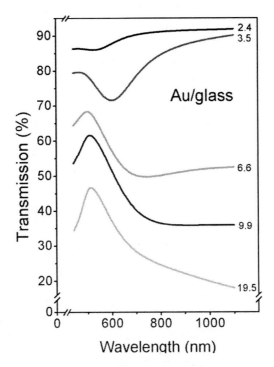

Figure 14. Transmission spectra of Au layers on glass substrate taken for different layer thicknesses (in nm) as indicated in the figure [71].

Evaluations of optical spectra were performed using Film Wizard software and a Maxwell-Garvett model was applied. In this model Au films is described as a heterogeneous mixture of material and voids. With the aim to incorporate nanosize of gold clusters for above mentioned material was presumed the Lorentz-Drude behaviour of the optical parameters. This approximation is generalization of both Lorentz oscillator and Lorentz-Drude models and include the effect of the free-carrier contribution to the dielectric function and resonant transitions between allowed states. The best fits were obtained in the case of thickness from 2-15 nm. Main parameters of the chosen approximation - plasma frequency and center oscillator energy are presented at the figure 15 as a function of the film thickness. Center oscillator energy could be attributed to optical excitation by surface plasmon excitation. As it was predicted by the theory of Mie [89], the red shift appears with increasing of cluster size (film thickness). Additionally, it is evident, that plasma frequency strongly depends on the film thickness. The plasma frequency increases with increasing layer thickness and for thicknesses above 15 nm it reaches typical "bulk" value of the gold - 9.02 eV. It is well known, that the plasma frequency is closely related to the concentration of free carrier. From figure 8 it can be concluded that the concentration of free carriers is an increasing function of the film thickness. This result is in good agreement with previous studies [79].

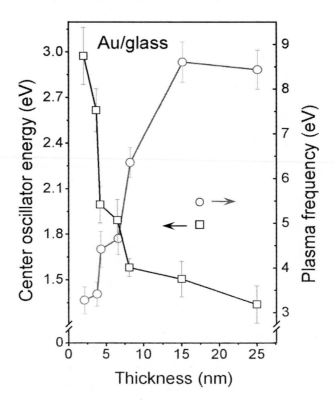

Figure 15. Center oscillator energy (□) and plasma frequency (○) calculated from optical spectra as a function of Au layers thickness [71].

3.4. LATTICE PARAMETER AND DENSITY OF AU LAYERS

It has been published elsewhere [52,63,71-75], that lattice parameter of metals prepared in the form of thin layer by a physical deposition is not a material constant but depends strongly on the layer thickness. Figure 2B displays the dependence of the Au lattice parameter on the layer thickness derived from the present XRD measurements. The lattice parameter monotonously declines of with increasing layer thickness. The decline can be explained by the internal stress relaxation during the growth of gold clusters (see [63]).

With the aim to find how the decline of the lattice parameter influences the density of the gold layers, the effective thickness and the mass of the deposited layers were measured and the effective density calculated in a standard way. In figure 16 (curve 1) the dependence of the density on the layer thickness is shown. The density increases with increasing layer thickness and for about 90 nm thick layer it achieves the density of bulk gold. The reduced density of thinner layers is probably due to higher fraction of the free volume in gold nano-clusters. As the gold clusters become greater [1], the free volume fraction decreases and the gold density gradually increases. It was reported earlier [90] that the gold layers with thicknesses above 100 nm prepared on glass substrate, exhibits quite uniform density, with mean value of 19.320 g cm^{-3} typical for bulk material.

In figure 16 the measured densities (curve 1) are compared with those calculated for extreme lattice parameters a = 0.40835 and 0.40731 nm (curves 2 and 3 respectively). The densities of Au layers were calculated from the relation $\rho_{Au}=m_{Au}/V_{nanoparticle}=(N_{Au}.M_{Au})/(V_{Au}+V_{free})$, where m_{Au} is weight of all Au N_{Au} atoms presented in nanoparticle block, $V_{nanoparticle}$ is volume of nanoparticle block including the Au atoms volume V_{Au} and free volume V_{free} (the volume between individual atoms in nanoparticle block), M_{Au} is a mass of one Au atom. $V_{nanoparticle}$ is obtained from a lattice parameter a (determined by XRD) and a diameter of Au atom r_{Au}=0.144 nm.

Density calculated from the lattice parameters exceeds that obtained from effective thickness and mass for all layer thicknesses considered. In fact, curves 2 and 3 (see figure 16) were calculated for a single cluster covering whole sample surface. In the case of the curve 1 the value of the effective thickness incorporates the influence of free volume and non-covered places (space between individual islands) into the density determination. Thus the dissimilarity of curves 1 and 2, 3 can satisfactorily be explained either by the influence of free-volume fraction or of non-covered surface during initial phases of the layer growth [1]. Obviously, in the initial phase of the layer growth the difference between the lattice density of a cluster and the effective density of the layer is most pronounced. As the layer become more homogeneous, the difference gradually disappears and both densities tend to the value of bulk gold.

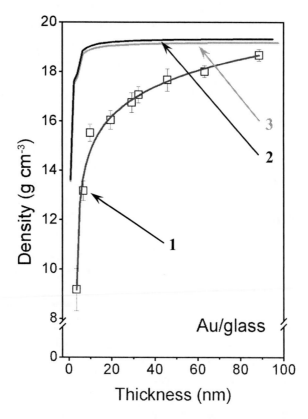

Figure 16. Dependence of Au density on the thickness of Au layer deposited on glass substrate. The densities were calculated from effective thickness and weight of Au layers (curve 1) and calculated from two border values of lattice parameters (a = 0.40835 and 0.40731 nm, curves 2 and 3) [71].

3.5. SURFACE MORPHOLOGY AND ROUGHNESS

The images that illustrate surface morphology and roughness (R_a) determined from AFM measurements of pristine and gold coated PET and PTFE are shown in figure 17a [79]. It is clear that pristine PET and PTFE exhibit different surface morphology and roughness, the difference being due to different fabrication technique used for oriented PET and cut PTFE films. After deposition of 20 nm thick gold layer onto PET the surface roughness increases and the surface morphology changes. In contrast, the same deposition onto PTFE does not result in any significant changes in the surface morphology and roughness. The effects of Au deposition on surface morphology and roughness of Au/polymer structure will be discussed in next chapter.

Figure 17a. AFM scans of pristine PET (PET), PET sputtered with 20 nm thick Au layer (PET/Au), pristine PTFE (PTFE) and PTFE sputtered with 20 nm thick Au layer (PTFE/Au). Ra is average surface roughness in nm [16].

We have studied various physical and physico-chemical properties of gold nano-structures deposited on different substrates (glass, silicon and some polymers) [32,33,71]. In work [71a] gold structures were sputtered onto glass. Then the samples prepared in this way were annealed at 300°C for one hour. The effects of annealing on gold structures sputtered onto glass were studied using AFM (surface morphology and roughness), UV-Vis (optical band gap) techniques and electrical measurements (sheet resistance) [71a].

The AFM images that illustrate the surface morphology and roughness (R_a) of gold coated glass before and after annealing are shown in figure 17b. For the sake of comparison only images from the samples with identical vertical scale were chosen. From figure 17b it is clear that the surface morphology of the sputtered structures does not depend significantly on the sputtering time. After annealing, however, the surface morphology changes dramatically. It is seen from figure 17b that the annealing leads to formation of „spherolytic and hummock-like" structures in the gold layers. The formation may be connected with an enhanced diffusion of gold particles at elevated temperature and their aggregation into larger structures. It is well known that the melting point of the gold nano-particles decreases rapidly with decreasing particle size [54]. The migration of the gold nano-particles and formation of larger structures may be connected with lower thermodynamic stability of the gold nano-particles and lower gold wettability of glass. This idea is supposed by some previous XRD experiments in which dominant (111) orientation of gold crystals in the sputtered gold layers was determined. The (111) oriented gold crystals are known to be thermodynamically unstable and their melting and cracking starts from the edge parts that should be bounded to Au (110) surface [1,54].

On sputtered samples the darkening of the deposited structure and changes of colour from blue to green, both connected with increasing structure's thickness were observed. After annealing the „same structures" acquire red colour. Non-zero band gap indicates semi-conducting properties of sputtered structures. Annealing leads to significant increase of the optical band gap. For sputtered gold structures the rapid decline of the sheet resistance R_s is observed for sputtering times above 50 s. For the annealed structures the decline appears on the structures deposited for the times above 250 s. Annealing also results in dramatic changes in the gold layer morfology and surface roughness. According to AFM images „spherolytic and hummock-like" structures are created in the gold structures by annealing [71a].

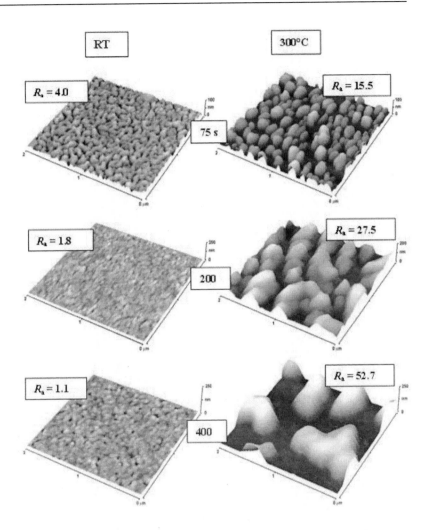

Figure 17b. AFM scans of gold structures sputtered for 75, 200 and 400 s on glass substrate before (RT) and after annealing (300°C). R_a is average surface roughness in nm [71a].

Chapter 4

GOLD LAYERS EVAPORATED
OR SPUTTERED ON POLYMER

4.1. AU LAYER DEPOSITION ON PRISTINE POLYMER

Metal layer can be prepared on the polymer surface by sputtering or vacuum evaporation [16]. Microscopic theory of nucleation processes for sputtering and vacuum evaporation was suggested by Walton [91] and for polymer–metal systems by Thran et al. [92]. Two nucleation mechanisms are usually considered. In so-called preferred nucleation, metal atoms are trapped at preferred sites while, in random nucleation, nuclei are formed by metal atom encounters. Both processes have been observed in polymer metallization [92].

The continuity, homogeneity, and surface morphology of Au layers prepared by sputtering on polymer surface were studied earlier [80,93]. Au layers prepared by sputtering and vacuum evaporation on PET substrate are characterized and their parameters (see [2]). In figure 18, the dependence of the layer thickness on the deposition time is shown for both deposition techniques. The layer thickness, calculated from Au amount detached from the sample surface and determined using AAS method, represents an average layer thickness. Because of limited sensitivity of the AAS method, it was impossible to measure the layer thickness for „very thin" layers. It is obvious that the sputtering proceeds with two different deposition rates given by the slope of the layer thickness versus deposition time dependence. The deposition using vacuum evaporation proceeds with a constant deposition rate. To facilitate the comparison of both deposition techniques in transient region between continuous and discontinuous layers, the deposition rate of vacuum evaporation was roughly normalized to that of sputtering for shortest deposition times.

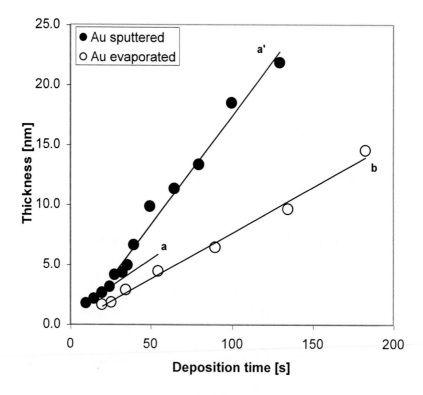

Figure 18. Dependence of the Au layer thickness deposited on PET substrate on deposition time for sputtered (a, a′) and evaporated (b) Au layers [16].

The measured sheet resistance R_s (A) and 1/vswr values (inverse voltage standing waves ratio) (B) as a function of the layer thickness are shown in figure 19 [2]. Both complementary quantities well characterize the transition from discontinuous to continuous, homogenous metal coverage. Extremely rapid resistance decrease at a critical layer thickness, indicating transition to continuous layer, is observed on sputtered Au layers (figure 19A). For vacuum evaporation much slower decrease of R_s is observed and continuity is achieved for larger layer thickness. The difference can be explained by different mechanisms of Au deposition, deposition of separated atoms by sputtering and larger atomic clusters by vacuum evaporation. The deposition of separated atoms facilitates creation of continuous layer.

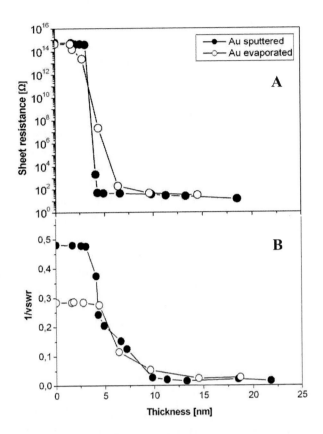

Figure 19. Dependence of the sheet resistance (A) and value 1/vswr (B) on layer thickness for sputtered and evaporated Au layers on PET substrate [16].

From the comparison of figure 19A and B it is obvious that the quantity 1/vswr, characterizing reflection of electromagnetic waves, is less sensitive to the layer continuity in comparison with sheet resistance. This can be explained by the fact that the electromagnetic wave reflection depends not only on the layer continuity but also on its homogeneity. According to the sheet resistance measurement, the layer continuity is achieved at lower layer thickness for sputtered layers in comparison with evaporated ones. However, there is no principal difference, within experimental errors, between electromagnetic wave reflection from the continuous layers prepared by both methods. According to the measured 1/vswr values, both sputtered and evaporated layers are discontinuous for thicknesses below 4 nm, continuous but heterogeneous for thicknesses from 4 to 10 nm, and continuous and homogenous for thicknesses above 10 nm.

Surface morphology of deposited Au layers was characterized [16] by AFM (figure 20) and SEM (figure 21) techniques. The differences in the surface morphology of layers with similar thickness prepared by sputtering and vacuum evaporation are obvious. It is seen from figure 20 that oriented PET film exhibits homogenous surface, the roughness of which was about 1 nm according to AFM measurement. As expected, the Au deposition leads to changes in the surface morphology and roughness in comparison with pristine PET. On the layers prepared by vacuum evaporation no significant changes in the surface morphology are observed in dependence on the layer thickness (figure 20). Rounded Au clusters appears on the sample surface. Most pronounced roughness along the z-coordinate was observed on the 4.9 nm thick Au layer. Quite another surface morphology was observed on the layers prepared by sputtering. Au in discontinuous layers creates much smaller, clusters. With increasing layer thickness, the clusters enlarge and at the same time the surface roughness increases. The differences in the surface morphology can be explained by different mechanisms of Au deposition, single atoms in sputtering and atomic clusters in vacuum evaporation. Needle-like surface of sputtered, discontinuous layers may be due to the mechanism of Au deposition, in which the initial metal islands serves as preferential nucleation centers for coming Au atoms. This is not the case for dissimilar polar substrates [92].

The SEM images of the same samples as in figure 20 are presented in figure 21. SEM imaging of pristine PET is impossible because of sample charging. This effect is also observed on samples with discontinuous Au layers (thickness about 3 nm) prepared by both techniques and partly also on the evaporated layer 4.9 nm thick. As to continuity of the deposited layers, the results of SEM imaging are in accord with those obtained in the measurement of sheet resistance and reflection of electromagnetic waves (figure 19). By comparison with figure 18, it is obvious that for the thicknesses above 5 nm the layer is continuous, but in SEM imaging an island-like structure is still observed. In contrast to AFM results, in the SEM imaging, no significant differences are observed in worm-like structure of layers prepared by sputtering and vacuum evaporation. With increasing layer thickness, the surface regions in which the Au presence is not detected by SEM diminish. The present results confirm the well known fact that for the SEM imaging of insulators the Au coverage of samples by vacuum evaporation is more suitable than by sputtering, since the former technique leads to smaller deformation of the sample surface.

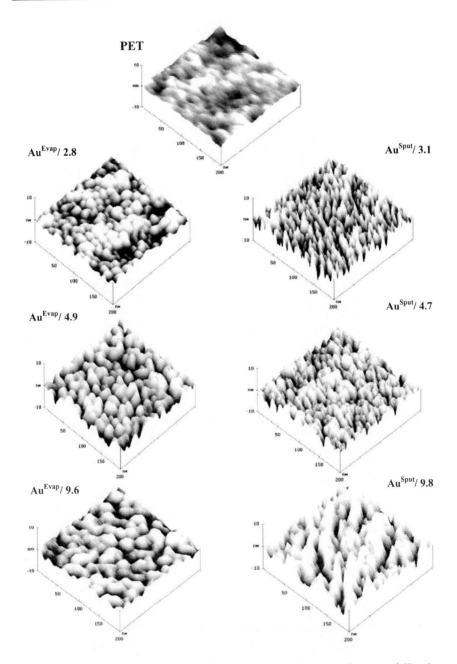

Figure 20. AFM images of pristine PET and evaporated (Evap) and sputtered (Sput) Au layers on PET. The numbers denote layer thickness of deposited Au layers in nanometers [16].

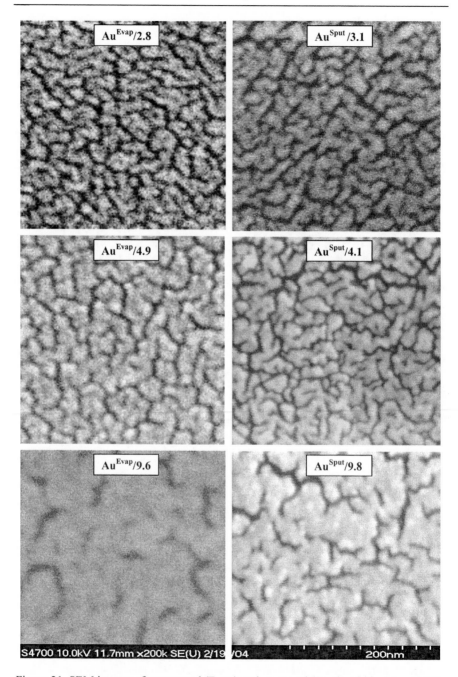

Figure 21. SEM images of evaporated (Evap) and sputtered (Sput) gold layers on PET. The numbers denote layer thickness of deposited Au layers in nanometers [16].

4.2. AU LAYER DEPOSITION
ON PLASMA TREATED POLYMER

Gold layers of different thicknesses were evaporated on pristine and plasma modified polymers. Surface morphology, gold layer thickness, and electrical resistance of structures prepared under different conditions were determined [94]. Wettability of plasma-modified PET and PTFE samples was determined using the goniometric method with water drops. Dependence of the contact angle on the time elapsed from plasma treatment is shown in figure 22. The plasma treatment of polymers leads to a dramatic decrease of the contact angle compared to pristine polymers, which is a well-known phenomenon in the field of plasma modification [95,96]. Gradual growth of the contact angle after the plasma treatment is a result of spontaneous rearrangement of degraded macromolecules and molecular fragments in the polymer surface layer (e.g. reorientation of polar groups toward the polymer interior) [95,97]. The relaxation processes are affected by interaction with ambient atmosphere during the aging process.

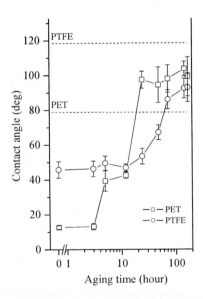

Figure 22. The dependence of the contact angle of plasma-modified PET and PTFE on the time elapsed from the plasma treatment. Dashed lines denoted by symbols PET and PTFE show the contact angle of pristine polymers [94].

In order to specify the compositional and structural changes in the polymer surface layer as a function of the aging time, the content of oxygen (PET, PTFE) and fluorine (PTFE) was determined using the XPS method applied immediately after the plasma treatment and after 168 h of aging (Table 1). In pristine PET, declared oxygen content is 28.6 at.%; the content of hydrogen was neglected because the XPS method is not able to measure H directly. Surprisingly, much less of oxygen and fluorine in pristine PET and PTFE surface layers is observed in comparison with what was expected. It may be speculated that an orientation of the oxidized structures toward the polymer bulk already takes place during fabrication of the oriented polymer foil. One can see from Table 1 that the plasma treatment results in a pronounced oxidation of the PET and PTFE surfaces. Decrease of fluorine content in PTFE indicates that the C–F bond is also degraded by plasma treatment. Aging for 168 h leads to a decrease of initial oxygen content in both polymers, indicating a decline in the polar group content and correlating well with observed behaviour of the measured contact angle (figure 22).

**Table 1. Atomic concentration of oxygen in PET and PTFE
and fluorine in PTFE, as calculated from chemical formula (theory)
and those determined by XPS method in surface monolayer of
pristine polymers (pristine), plasma modified polymers measured
1 h after the treatment (new), and the same samples
measured 168 h after the treatment (old) [94]**

Atomic content of oxygen and fluorine (at.%)			
Sample	PET	PTFE	
	oxygen	oxygen	fluorine
theory	28.6	0	66.7
pristine	2.4	0	62.5
modified (new)	59.2	6.8	55.1
modified (old)	37.8	5.2	51.4

Another factor influencing the wettability of the polymer is its surface morphology. Roughness of both pristine and modified PTFE is much higher than that of PET because of different manufacturing methods of the polymer foils (PET was oriented, PTFE was sliced) [94]. It is seen that the roughness increases as a result of the plasma treatment. This result can be explained by a preferential ablation of the amorphous polymer phase and a corresponding increase of the crystalline phase [43]. This supposition is supported by study

[98] in which crystalline regions in the form of spherulites were observed on the PET surface ablated by an excimer laser. After 168 h elapsed from plasma treatment, a mild decrease of surface roughness is observed. This decrease is probably due to a reorientation of polymer segments in the polymer surface layer.

The Au layers were deposited on pristine and plasma modified PET and PTFE. The deposition on the plasma modified polymers was accomplished immediately after the plasma treatment (new samples) and after 4-week aging period (old samples). The results of thickness measurements are summarized in Table 2. The aim was to examine how the surface structure and character of pristine and modified polymers affect the thickness and morphology of the deposited Au layer. The layer thickness on the polymer substrate was measured by the AAS-based technique and that deposited on the control Si sample was measured by a profilometer. The AAS data were converted into layer thickness using, deliberately, bulk Au density, and so they are systematically lower than that determined by the profilometer. It is seen that, under the same deposition conditions, the layers deposited onto pristine PTFE are much thinner. The thickness of the Au layers deposited onto pristine PET and both plasma-modified polymers are, within experimental errors, the same.

Table 2. Thickness of vacuum-evaporated Au layers on PET and PTFE substrates, determined by the AAS method on pristine substrate, and substrate modified by plasma discharge. The values are given with a typical uncertainty of ±10%. The thickness of the layer deposited on Si backing was measured by a profilometer. The sample denotation is the same as in Table 1 [94]

Thickness of evaporated Au layers (nm)						
Si	PET			PTFE		
	pristine	new	old	pristine	new	old
80 ± 3	56	63	57	21	62	60
50 ± 2	32	30	32	7	34	36
20 ± 7	10	11	10	2	11	10

The sheet resistance (R_s) of Au layers deposited on pristine and plasma-modified polymers as a function of the layer thickness (measured on Si substrate) is shown in figure 23. The resistance decreases with increasing layer thickness on all samples [80]. The layers deposited on pristine PTFE, with the thickness of 2 and 7 nm (according to AAS data, 50 nm measured on Si), are supposed to be discontinuous [80], and sheet resistance could not be measured on these layers because the upper limit of the device was 50 MΩ. A more pronounced decrease of the sheet resistance with increasing layer thickness was observed on plasma-modified PTFE than on PET samples. Repeated measurements show that the sheet resistance remains unchanged, within experimental errors, during 2 months of aging.

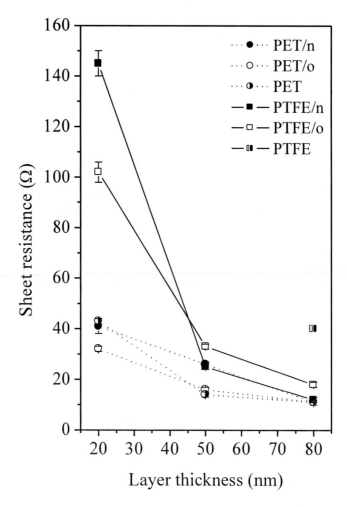

Figure 23. Sheet resistance as a function of the Au layer thickness deposited on PET and PTFE samples. The resistance was measured on Au layers deposited on pristine polymers, on modified polymers immediately after the plasma treatment (n-new), and after 4 weeks of aging (o-old) [94].

The results of AFM examination of the surface morphology of 50 nm thick Au layers deposited on pristine and plasma modified PET and PTFE are illustrated in figures 24 and 25. It is seen that Au deposited on the modified PET creates larger agglomerates than that deposited on pristine PET. More pronounced agglomeration leads to higher roughness R_a of plasma-modified PET samples (figure 24). More significant differences in the surface morphology and roughness are observed between Au deposited on pristine and plasma-modified PTFE (figure 25). It is seen that, in comparison with pristine PTFE, the deposited Au creates larger agglomerates and the layer exhibits a higher roughness on plasma-modified PTFE substrate [94].

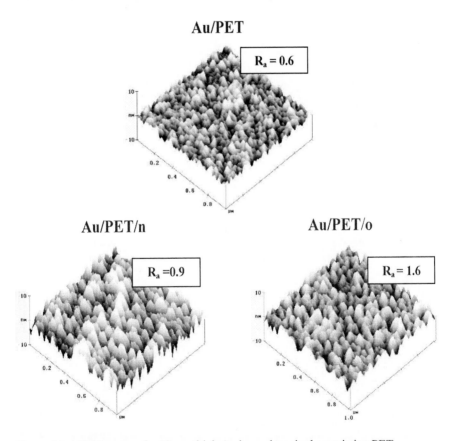

Figure 24. AFM images of a 50 nm thick Au layer deposited on pristine PET (Au/PET), modified polymer immediately after the plasma treatment (Au/PET/n-new), and after 4 weeks of aging (Au/PET/o-old). The Ra value characterizes the surface roughness in nm [94].

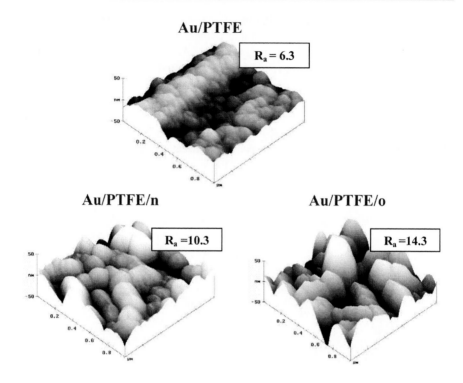

Figure 25. AFM images of a 50 nm thick Au layer deposited on pristine PTFE (Au/PTFE), modified polymer immediately after the plasma treatment (Au/PTFE/n-new), and after 4 weeks of aging (Au/PTFE/o-old). The R_a value characterizes the surface roughness in nm [94].

Chapter 5

GOLD NANO-WIRE ON POLYMER SUBSTRATE PREPARATION

Exposure to plasma discharge or irradiation with laser light affect physical and biological properties of polymers [99] and adhesion of subsequently deposited metal layers [100,101]. These effects are due to the changes in polymer surface morphology and chemistry, e.g. as a result of formation of oxygen containing groups [100-102]. Nano-structured materials and metal nanolayers are being extensively studied in microelectronics as basic components of sensors, actuators, photovoltaics, polymeric displays [103,104].

Several studies have shown that the illumination of polymers by polarized UV laser beam induces self-organized ripple structure formation within a narrow fluence range well below the ablation threshold [105,106]. The properties of these periodic structures have been frequently studied since their first observation [107]. So far the smallest structures, produced in this way, were created by our group by irradiation of polyethyleneterephthalate with 157 nm F_2 laser [33,108].

The period of the ripples depends on the laser wavelength and on the angle of incidence of the radiation, and their direction is related to the laser beam polarization [109]. The spacing of the ripples can be described by $\Lambda = \lambda / (n - \sin \theta)$ (1), where Λ is the lateral periodicity of the ripples, λ is the wavelength of the excitation laser light, n is the effective refractive index of the material and θ is the angle of incidence of the laser beam [109].

These structures are formed by irradiation with one laser beam with uniform intensity distribution of material with small initial roughness. It is known that the interference between the incoming and the surface scattered waves plays an important role in the structure formation [110]. The interference causes an inhomogeneous intensity distribution, which together with a feedback mechanism results in the enhancement of the modulation depth [111]. However, the whole mechanism is complex and different processes have been reported as responsible for ripple formation [108,112] such as thermal and non-thermal scissoring of polymer chains, amorphization of crystalline domains, local surface melting, photo-oxidation and material transport.

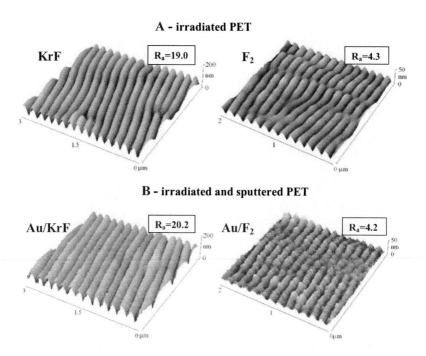

Figure 26. AFM images of PET samples: (A) irradiated by KrF (6.6 mJ cm^{-2}) and F$_2$ (4.4 mJ cm^{-2}) lasers and (B) irradiated and than sputtered with 200 nm (KrF) and 50 nm (F$_2$) Au layer. R_a is average surface roughness in nm [113].

The laser induced ripples have a fluence-independent width Λ, which is given by formula (1), while the height of the structures strongly depends on the fluence laser light. Further experiments were done on the samples with the most pronounced changes in the surface morphology, i.e. with the highest ripple structures [113]. In figure 26, the surface morphology of the samples irradiated with F$_2$ (fluence 4.4 mJ cm^{-2}) and KrF (fluence 6.6 mJ cm^{-2}) lasers is compared. The ripples formed by exposure to KrF laser light have larger width and height in comparison to those produced by F$_2$ laser light. The periodicity,

Λ, of ripples was about 208 nm and 140 nm for irradiations with KrF and F_2 laser respectively. The height of the ripple structure (top-bottom) was about 100 nm for KrF laser irradiated sample and about 15 nm for the F_2 laser irradiation. In the next experiment gold layers it was sputtered onto the ripple structures and their morphology was studied. The ratio of the gold layer thickness to ripple height was kept constant for the sake of comparison. [113]. Figure 26 shows morphology of gold layers sputtered onto laser-induced ripple structures for both laser wavelengths mentioned above. For both wavelengths, the ripple structure on the PET surface is imprinted on the gold layers. The roughness values R_a were nearly identical for the surface with and without gold coating in both cases. However, for the F_2 laser irradiated samples it seems that the gold coated surface has a more pronounced granular structure than the uncoated ripple structure. Here one has to keep in mind that the AFM images show a convolution of the surface morphology with the geometry of the AFM tip. Therefore, small features and features with a high aspect ratio may not be feasible in the AFM images.

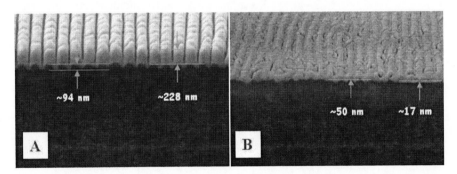

Figure 27. FIB-SEM images of lasers irradiated and Au sputtered samples: (A) PET irradiated by KrF laser (6.6 mJ cm-2) and sputtered with 200 nm of Au layer, (B) PET irradiated by F2 laser (4.4 mJ cm-2) and sputtered with 50 nm of Au layer [113].

Supplementary to the AFM analysis, the samples were cut by FIB. The FIB cuts were then investigated with the SEM (figure 27). The gold sputtered onto the KrF laser induced ripples is deposited in the form of "nano-wires", which grow on the ridges of the ripples. SEM images of the cuts indicate that the metal layer is not continuous and that there are gaps between the individual wires. Additionally, a kind of graining seems to be visible along the wires, but the FIB cut images suggest that these grains are connected each other. The morphology of the gold layers deposited onto the F_2 laser induced ripples is quite different. The gold is also deposited in the valleys of the ripple structure and the ripple profile is partially smeared out. The layers are continuous but they show cracks which are not directly related to the ripple pattern.

Chemical composition of the polymer surface can affect the growth and morphology of gold layers on polymer surface [94]. Therefore, we analyzed KrF and F2 irradiated PET with the ARXPS method [114]. ARXPS measurements show that oxygen concentration in pristine PET is depth dependent especially in the surface layer about ten atomic monolayers thick [114]. This finding is explained by a reorientation of the polar groups in the PET surface layers resulting in lower surface concentration of oxygen in comparison with expected value for pristine PET (29 at.%) [115]. The dependence of the oxygen concentration on the angle of incidence of the XPS primary beam is shown in figure 28 for pristine PET irradiated with PET KrF and F_2 lasers. For pristine PET the oxygen content decreases from 25.5 at.% for the incidence angle of 0° to 19.5 at.% for the incidence angle of 80°. For PET irradiated with F_2 laser the oxygen concentration declines from 18 at.% to about 16 at.% and the PET irradiated with KrF laser the oxygen concentration increases from 28.5 to 33 at.% (the values obtained for incidence angles 0 and 80°).

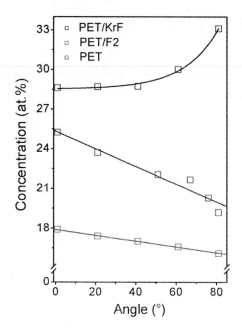

Figure 28. Dependence of the oxygen concentration (at.%) on the angle of incidence of the XPS primary beam for pristine PET and PET irradiated by KrF and F_2 lasers [113].

Since the larger incidence angles correspond to smaller penetration depth, we conclude that the oxygen content remains nearly unchanged in the upper about 10 nm thick layer of the irradiated PET surface, which can typically be examined by the XPS method. From this we conclude, that the ridges and the valleys of the F_2 laser-induced ripples have nearly similar chemical composition, at least regarding the oxygen content. Surprisingly the oxygen content in F_2 irradiated PET is lower than in pristine PET even though the surface is modified, meanwhile in the case of KrF laser irradiation the oxygen content is higher and towards to the surface the oxygen content increases. This implicates that in the case of the KrF irradiation the degradation is more pronounced compared to the F_2 laser irradiation. In case of the KrF laser irradiation the ridges of the ripple structure probably contain more oxygen than the valleys. The oxygen concentration on the ridges is even higher than that in pristine PET. This could be due to an enhanced degradation of the polymer on the ridges during the laser irradiation.

The oxygen concentration on the PET surface irradiated with the F_2 laser is lower than that in pristine PET. The effect may be related to preferential reorientation of the polar groups toward the polymer interior [114,115] or by photochemical scission of the polymer chains in combination with preferential release of CO and CO_2 groups (Norrish type II reaction, [116])

Chapter 6

ADHESION OF AU LAYERS
ON POLYMER SUBSTRATE

6.1. MODIFICATION OF POLYMER SURFACE
BY PLASMA DISCHARGE

It was observed elsewhere that the plasma treatment of polymer results in cleavage of macromolecules, alterations of chemical structure and ablation of surface layers, and thus affects surface properties of polymer e.g. its solubility [117]. The first objective of the chapter is to summarize and discuss the effects of the plasma treatment of polymer.

The thickness of the ablated layer of the plasma-treated polyethylene (PE), PET, polystyrene (PS), and PTFE measured by gravimetry is shown in figure 29 [118]. One can see that similar ablation is observed on PE and PET while on PS markedly thinner layer is ablated. Under the same experimental conditions about 50 nm thick surface layer is ablated from PTFE. The high ablation rate of PTFE is well known and it is probably due to the presence of highly reactive fluorine radicals in the plasma [119].

The surface morphology of polymers was examined by AFM microscopy (see [118]). Obviously, the plasma treatment leads to changes of surface morphology of all polymers. The roughness of pristine PTFE (figure 30) is more than 25 times greater than that of, for example, pristine PET. This is due to different manufacture of both polymer foils; PET is oriented and PTFE is sliced. Pronounced changes of the PTFE surface morphology after the plasma treatment is observed [118]. The 'worm-like' structure of pristine PTFE disappears after the treatment and a considerably diverse and 'hummock' relief with occasionally 'dimpling' arises. The surface morphology of the treated PTFE completely differs from that of the pristine polymer and the surface roughness increases (figure 30). It may be concluded that the pronounced dissimilarity of pristine and plasma-treated PTFE surfaces is caused by enormous ablation. According to the earlier data reported in [120], the amorphous phase of PTFE is ablated at a faster rate than the crystalline one.

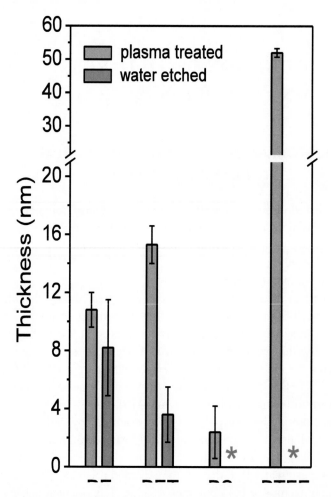

Figure 29. Thicknesses of the ablated polymer layers and layers dissolved by water etching measured by gravimetry (* denotes immeasurable value) [118].

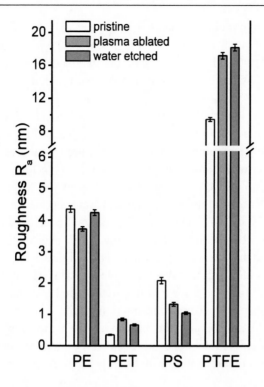

Figure 30. Roughness R_a of pristine, plasma-treated, and plasma-treated and water-etched PE, PET, PS, and PTFE [118].

It may be concluded that plasma treatment leads to the ablation and surface morphology changes which are different for the different polymers investigated. The surface roughness of PET and PTFE increases after the plasma treatment, while it decreases in the case of PE and PS. Since the changes in the surface morphology may enhance the polymer interaction with polar solvents [121,122], the water etching of modified polymers was examined in the next experiment. The thicknesses of the surface layers removed from plasma-treated polymers by 24 h water etching and measured by gravimetry are shown in figure 29. For PS and PTFE no data are presented since the weight loss was negligible. For PET and PE measurable weight loss was observed after the water etching. The thickness of the surface layers removed from plasma-modified PET and low-density polyethylene (LDPE) by water etching is about 4 and 8 nm, respectively. The comparison of the plasma-treated and the plasma-treated and subsequently water etched samples shows that tiny, sharp formations (LMWO segments) are removed from the sample surface by the etching. The etched surfaces exhibit lower diversity in some cases. No obvious tendency in roughness evolution is observed (see figure 30). As a result of the water etching the PET and PS surface roughness slightly decreases, while an increase of the surface roughness is observed for LDPE and PTFE.

The chemical structure of plasma treated PE was examined using XPS, the oxygen concentration profile was determined from Rutherford Backscattering Spectroscopy (RBS) measurement and the concentration of free radicals was determined by Electron Paramagnetic Resonance (EPR) technique [6]. So far -CH a $-CH_2$ (of pristine PE), oxidized (-C=O, -COO,-COC-), $-NH_2$, and also -C=C- groups (typical for aromatic substances) were reported to be present at Ar plasma treated HDPE [43]. Depth profile and total content of oxygen in PE was determined by RBS method. The profiles of plasma treated sample and that subsequently etched in water are compared in figure 31 [123]. The mean oxygen contents in the surface layer accessible by RBS, ca 140 nm thick, are (39 ± 5) and $(28\pm3).10^{15}$ cm^{-2} for plasma treated and plasma treated-water etched samples respectively. Both concentration profiles exhibit a maximum at the depth of ca 20 nm, behind the maximum the concentration decreases slowly, falling to negligible concentrations at 60 nm. Similar trend was observed in previous report dealing with PE exposed to Ar plasma [43].

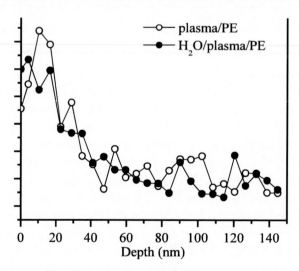

Figure 31. Concentration depth profiles of oxygen obtained from RBS measurement of plasma treated PE and PE plasma treated and water etched for 24 hours [123].

Due to plasma treatment the radicals (free spin) are generated on polymer chain. Not only C-H but also C-C bonds are likely to break, the later leading to fragmentation of polymer chain. The radical density determined by EPR method is in Table 3. The density of free radicals decreases during the aging to about quarter of the initial after 80 days. The decrease of radical density is a result of radical recombination [124].

Table 3. C (1s) a O (1s) concentrations (at.%) determined from XPS measurements and radical densities (10^{18} g^{-1}) determined by EPR method. The concentrations and densities were determined on the pristine PE, plasma treated PE (1 or 24 hours after the treatment) and plasma treated and water etched PE (see left hand side of the table). Before the measurement the samples were stored under standard laboratory conditions [114]

PE modification	C	O	Radical density
pristine PE	100	0	-
1 h after plasma treatment	69.2	30.8	-
24 h after plasma treatment	75.7	24.3	2.77
plasma treatment and 24 h in water	63.4	36.6	2.31

Surface wettability was determined by measuring the water contact angles. It was observed earlier that the wettability depends on the time the sample spends in contact with the atmosphere after the exposure to the plasma discharge [125]. Figure 32 shows the contact angle as a function of the plasma exposure time and the time elapsed from the plasma treatment of PE. The water contact angle measured 5 min after the plasma treatment decreases from 70 (typical for pristine PET) to 35° for all samples. Then, with increasing aging time, the wettability decreases, the decrease being most significant for the samples plasma treated for more than 30 s. For these samples, the wettability reaches the saturation after 336 h. It is also seen that the contact angles of these samples are higher than those of the pristine PET. The saturation is probably due to complete rearrangement and relaxation of the degraded polymer chains and their fragments on the polymer surface [16,115] In this process, diffusion of light degradation products may play an important role too [97].

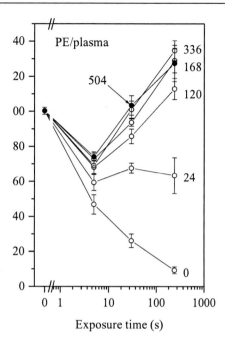

Figure 32. Dependence of the contact angle on the exposure time in 8.3 W plasma discharge. The contact angle was measured at different times (hours) elapsed from the plasma treatment (attached numbers) [125].

Using angle resolved XPS technique (ARXPS) [8], the chemical structure of the first ≈ 10 monolayers (about 5 nm in depth) of the sample surface can be characterized. The oxygen concentrations presented in figure 33 are calculated using standard statistical methods from the measured intensities and by using the attenuation factors quoted in ref. [126]. From the present XPS measurements, the atomic oxygen concentration in the surface layer was determined for both pristine and plasma treated PET samples. For plasma treated samples, the oxygen concentration was followed as a function of the aging time. The dependence of the oxygen concentration on the monolayer number counted from the sample surface, measured for the pristine PET and for the plasma treated one is shown in figure 33 as a function of the aging time. In all cases, the oxygen depth distribution is inhomogeneous. In the first monolayer of pristine PET the oxygen concentration is lower in comparison with the value expected from the PET stochiometry. The effect can be due to the preferential orientation of the oxygen containing groups inwards, which may occur in the process of fabrication of oriented PET foil. This seems to be in accord with higher oxygen concentration observed in deeper layers.

Immediately after the plasma treatment (PET new) the oxygen concentration increases, but then it decreases with increasing aging time (PET old). These results are in agreement with measurements of the contact angle (see above), the evolution of which is explained by the rearrangement of degraded macromolecules and molecular fragments.

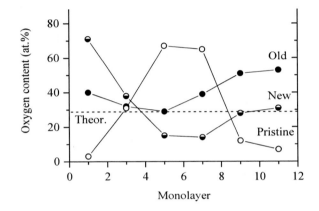

Figure 33. Oxygen concentrations measured by ARXPS on pristine and plasma treated PET. Plasma treated samples were measured after 1 h (New) and 1 week (Old) aging. Meanwhile the samples were stored under laboratory conditions (at RT and on the air). Theoretical value for the O concentration in the pristine PET (theor) is given for comparison [114].

6.2. AU LAYERS ADHESION ON POLYMER SUBSTRATE

Little information has been published concerning the interaction of gold with polymers.-Because of the lack of relevant information, we decided [114] to investigate the effect of Ar plasma treatment on the surface properties of PET and on metal-polymer adhesion. The samples were examined using a nanoindenter with the aim to determine mechanical properties of their surface [114]. Cross-linking creates a barrier for the diffusion of the deposited metal into the polymer "bulk", and in this way it may affect the nucleation and the properties of metal layer [127]. The microhardness of the gold coated PET is similar to that of pristine PET. Because of the measurement errors, it is not possible to prove unambiguously that the plasma modification and the aging of plasma modified PET have a significant effect on the microhardness of the Au layer (in contrast to significant effects on elastic modulus). The qualitative behaviour of the microhardness with the tip displacement is similar for all samples investigated [114]. Scratch tests were performed on Au coated pristine and plasma exposed PET samples. It is seen (figure 34) that there are no significant differences between the shapes of the single scratch paths, and no mechanical damages outside of plastic deformation is observed along the scratch path. It may be concluded that, in all cases presented here, the elastic deformation plays a main role. These results indicate that after the deposition of Au onto modified polymer surfaces, the Au atoms do not react with plasma activated sites (e.g. radicals), and therefore the plasma modification does not affect the adhesion of the Au layer. This is in contrast with the behaviour of more reactive metals such as Al and Cu [127,128]. These findings also differ from the earlier results we obtained under the similar conditions but using polyethylene as a substrate [129].

The main goal of our study [130] was to examine the effect of the plasma treatment and dithiol grafting on the binding of gold nano-particles and on the adhesion of sputtered gold layer on the PE surface (see figure 35). PE surface was modified by plasma discharge and subsequently grafted with 1,2-ethanedithiol. Short dithiol is expected to be fixed via one of –SH groups to radicals created by the preceding plasma treatment. Next, the free –SH group are allowed to interact either with gold nano-particles or with gold atoms from sputtered Au nano-layer.

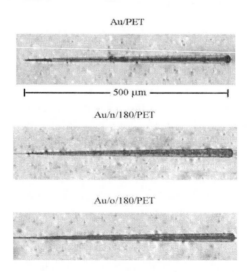

Figure 34. Optical images of the scratch paths produced on the PET surfaces by the movement of the tip gradually loaded from 0 to 20 mN [114].

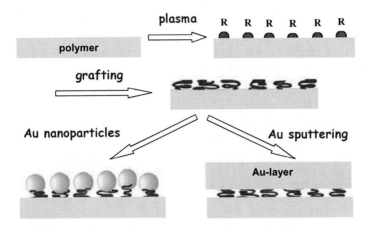

Figure 35. Scheme of polymer modification by plasma discharge (R-radical), grafting by dithiol and by either coating with Au nano-particles or by sputtering of Au nano-layer [130].

Radical density, determined from EPR measurement on the plasma modified PE (PE/plasma) and plasma modified and dithiol grafted PE (PE/plasma/SH), is shown in Table 4 as a function of the time from the plasma treatment. Plasma treatment results in scission of C-H and C-C bonds and production of free radicals (unpaired electrons). As could be expected, the radical concentration declines with increasing time from the treatment due to radical recombination [123]. Grafting of the plasma treated PE with dithiol leads to significant decrease of the radical concentration. The decrease indicates that the dithiol is chemically bound to the PE surface enriched with free radicals.

Table 4. Time dependence of radical concentration (10^{16} cm^{-2}) determined by EPR method for plasma modified PE (PE/plasma) and plasma modified and grafted with dithiol PE (PE/plasma/SH) [130]

Time after plasma treatment (hour)			
Sample	2	720	1800
PE/plasma	1.98	0.63	0.25
PE/plasma/SH	1.47	0.26	0.10

The surface composition of the PE samples modified by plasma discharge, dithiol grafted and coated with Au nano-particles was investigated using XPS and RBS methods [130]. By both methods the presence of chemically bound gold nano-particles is clearly proved. From XPS measurements the following surface concentrations were determined: C - 79.43, O - 19.43, Au - 0.50 and S - 0.82 at.%. Depth concentration profiles of O, S and Au were determined from RBS spectra. It is seen that the elements are present not only on the very surface, accessible to XPS, but they exhibit measurable depth profiles extending to the depth of about 250-300 nm [130]. The concentrations of all elements decrease monotonously with increasing depth. Minor difference between the depth profiles of Au and S may indicate that the smaller, sulphur containing dithiol molecules diffuse deeper than gold nano-particles (15 nm in diameter).

The mechanical properties of gold layers sputtered on pristine, plasma treated and plasma treated and dithiol grafted PE were examined using nanoindenter [130]. It is seen that the elastic modulus decreases rapidly from the surface up to the depth of 50 nm, which is the thickness of the sputtered gold layer. A plasma pre-treatment result in a dramatic decrease of the elastic modulus and another decrease is observed after the dithiol grafting. From the measured dependence of the hardness on the tip displacement it is seen that the plasma treatment results in a dramatic hardness decline. Subsequent deposition of Au, however, leads only to minor change in the hardness vs. displacement dependence.

The results of scratch tests are illustrated in figure 36. The scratch path in the sputtered gold layer for the pristine PE and plasma treated PE is qualitatively different (figures 36A,B). In several our studies [114,129] no significant effects of plasma treatment on the adhesion of sputtered gold layer has been observed. Dramatic changes in the shape and the character of the scratch path are observed on the dithiol grafted PE (see figure 36C). It is seen from figures 36 A-C that under the same indenter load (about 10 mN) most pronounced adhesive destruction of the gold layer takes place on the pristine PE, where underlying substrate appears. Lower adhesive destruction is observed on the gold layer deposited on the plasma treated PE (see figure 36B). On the dithiol grafted PE (figure 36C) adhesive destruction is much smaller and cohesive destruction becomes dominant. The cohesive destruction is demonstrated by radial spreading of scratches in the Au layer. It may therefore be concluded that the adhesion of the sputtered gold layer may greatly be enhanced by the dithiol grafting. The higher adhesion is probably due to chemical binding of the sputtered Au atoms to −SH groups present on the plasma treated and dithiol grafted PE.

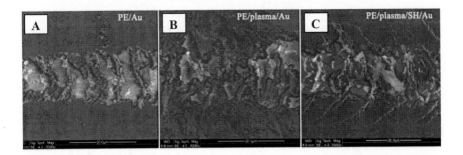

Figure 36. SEM images of the scratch path produced in Au layer, about 50 nm thick, deposited onto pristine PE ((A), PE/Au), plasma treated PE ((B), PE/plasma/Au) and plasma treated and dithiol grafted PE ((C), PE/plasma/SH/Au) [130].

Chapter 7

CELLS GROWTH ON AU NANO-PARTICLES GRAFTED ON POLYMER SUBSTRATE

The nanostructure of material surface (irregularities smaller than 100 nm) is considered to mimic the topography of the cell membrane, as well as the architecture of physiological extracellular matrix (ECM) molecules [40]. It is known that cell adhesion to artificial materials is mediated by ECM molecules spontaneously adsorbed to the material surface from biological fluids.

Various materials have been used for constructing nanostructured surfaces (e.g., polymers [131], ceramics, metals [40,132] and combinations of these materials). However, among metallic materials, gold, including its nano-sized forms, has shown no cytotoxicity in vitro and in vivo [47]. The non-toxicity of gold is related to its well-known stability, non-reactivity and bioinertness. For example, gold nanoparticles have often been conjugated with antibodies [133].

The hypothesis is that gold nano-particles may also serve as building blocks for creating artificial, bioinspired, nano-structured surfaces for tissue engineering. The plasma modified and gold grafted polymers with their nano-structured surface, appropriate chemical composition, wettability, electrical charge and conductivity may be an attractive substrate for cell colonization. In what follows the results of the study of adhesion and proliferation of rat VSMC and mouse fibroblasts of the line 3T3 on the modified PE surfaces is summarized [134].

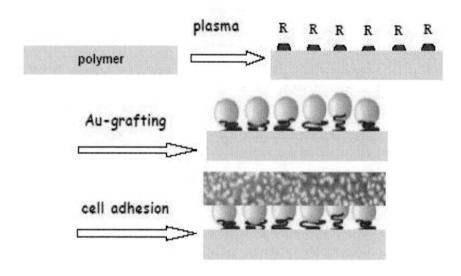

Figure 37. Scheme of polymer modification by plasma discharge (R-radical), grafting by dithiol and by either coating with Au nano-particles and adhesion and proliferation of cells.

Scheme of polymer modification by plasma discharge (R-radical), grafting by dithiol and by either coating with Au nano-particles and adhesion and proliferation of cell is shown in figure 37. Surface morphology of pristine PE and PE exposed to plasma discharge was examined using the AFM method (see figure 38). The plasma treatment led to an ablation of the PE surface layer [135]. As a result, the surface morphology was changed dramatically and surface roughness was increased. Also a lamellar structure appeared on the PE surface which could be due to the different ablation rates of amorphous and crystalline phases [43]. Further changes in the surface morphology and a decrease in the surface roughness occured after exposure of the samples to a water environment, the changes being probably due to preferential etching of short molecular fragments formed by the plasma treatment [118]. Figure 38 also demonstrates changes in the surface morphology and further decrease of the surface roughness after the grafting the plasma-activated and water-etched polymer surface with Au nanoparticles.

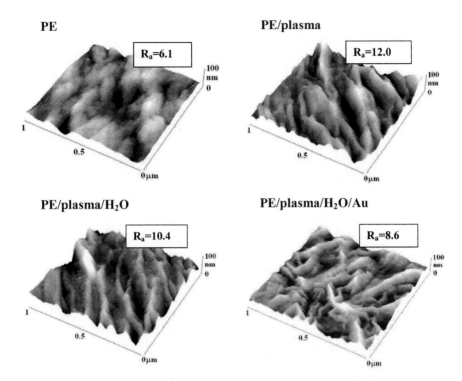

Figure 38. AFM images of pristine PE (PE), PE plasma treated for 150 s (PE/plasma), the same as before but subsequently etched in water (PE/plasma/H2O) and grafted with Au nanoparticles (PE/plasma/H2O/Au). Ra is the measured surface roughness in nm [134].

The presence of oxygen and nitrogen on the surface of the plasma treated and gold grafted PE was examined using the XPS method. The presence of oxygen on the polymer surface is seen from the XPS spectrum shown in figure 39 (O1s peak). Obviously, the oxygen from ambient residual atmosphere interacts with the plasma-activated PE surface, and creates different oxidized structures. Also, the Au4f peak was clearly seen indicating the presence of Au atoms on the surface of the gold grafted PE. The presence of gold was also proved by RBS measurement [134]. Au particles are present not only on the very surface of the PE, but also in the near surface layer about 100 nm thick and that the gold concentration decreases from the surface to the bulk [134]. The presence of gold particles beneath the PE surface may indicate an inward diffusion of gold particles, which could be facilitated by the structure defects created by the preceding plasma treatment.

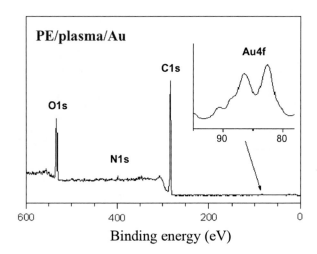

Figure 39. XPS spectrum of PE treated by plasma and grafted with Au nanoparticles. The O1s, C1s, N1s and Au4f peaks are clearly seen. In the inset, the enlarged region of the Au4f peak is shown [134].

The fact that no large Au agglomerates are observed on AFM images may support indirectly the concept of the gold particles penetration. The oxygen depth profiles simultaneously measured by RBS method and presented in our previous work [123] exhibit the same form as the Au depth profiles. Incorporated oxygen decorates the plasma modified surface layer. The measurement of the sheet resistance proved that the grafted Au particles did not create a continuous coverage of the PE surface.

It is well known that chemical structure and surface morphology significantly affect surface wettability [124,136], which in turn may affect adhesion and proliferation of living cells [124,136]. The contact angle, as a measure of the surface wettability, is shown in figure 40 like dependence on the plasma exposure time for plasma treated and gold grafted PE samples. For exposure times up to 50 s, the contact angle decreases from that measured on pristine PE to a minimum. Then the contact angle increases and for the samples exposed to the plasma discharge (for the times above 200 s), the contact angle exceeds that for pristine PE. A similar dependence is observed on the plasma treated and gold grafted samples, but the increase of the contact angle is slower and the contact angle remains below that on pristine PE. It may therefore be concluded that short time exposures to plasma discharge result in a rapid growth of hydrophilicity which can further be increased by the gold grafting. For longer exposure times, however, the hydrophilicity diminishes and finally the samples become hydrophobic.

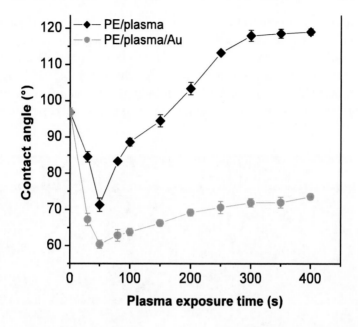

Figure 40. Dependence of the contact angle on the plasma exposure time for plasma treated PE (PE/plasma) and plasma treated and Au nanoparticle-grafted PE (PE/plasma/Au). The measurement was accomplished 20 days after the modification [134].

The cell-material interaction was studied on pristine PE, PE exposed to Ar plasma and plasma-activated samples grafted with Au nanoparticles [132]. It is apparent from figure 41 that the number of initially adhered VSMC is similar in all three groups of PE samples, i.e. this parameter is not influenced by treatment with plasma or by plasma treatment/Au grafting. However, both PE modifications enhance the proliferation activity of VSMC cells. Two days after seeding, the number of VSMC on both types of modified PE samples was higher than that on pristine PE samples. This difference increases in the following days of cultivation. On days 5 and 7 after the seeding, the number of cells was significantly higher on modified PE samples, especially on those grafted with Au nano-particles.

The higher proliferation activity of VSMC on modified PE samples can be explained by the oxidation of the polymer surface after the plasma treatment, resulting in the increase in the surface wettability. This increase was well apparent especially on the samples grafted with Au nano-particles, exhibiting increased wettability regardless of the plasma exposure time (see figure 40). It is believed that hydrophilic surfaces adsorb the cell adhesion-mediating molecules (such as vitronectin or fibronectin present in the serum supplement of cell culture media) in a more physiological conformation, which supports the accessibility of specific amino acid sequences (e.g. RGD) in these molecules to cell adhesion receptors (e.g. integrins) [45,137]. Receptor mediated adhesion and appropriate cell spreading on the adhesion substrate are prerequisites for proliferation capability in all types of anchorage-dependent cells, including VSMC [50,138].

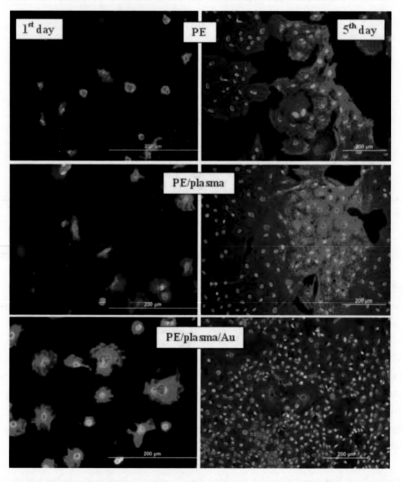

Figure 41. The photographs of VSMC adhered (1st day) and grown (5th day) on pristine PE (PE), PE treated for 150 s with plasma (PE/plasma) and subsequently grafted with Au nanoparticles (PE/plasma/Au) [134].

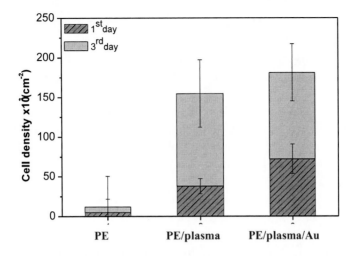

Figure 42. Dependence of the number of 3T3 cells on the cultivation time for pristine PE (PE), PE modified 150 s by plasma treatment (PE/plasma) and subsequently grafted with Au nanoparticles (PE/plasma/Au) [134].

Other factors influencing the growth activity of cells is the surface roughness and morphology of the substrate. Both plasma treatment and Au grafting create nano-scale irregularities on the material surface. The surface nano-roughness acts synergetically with the surface wettability, i.e. by improving the geometrical conformation of adsorbed ECM molecules, and thus a better exposure of active sites on these molecules to the cell adhesion receptors [40]. Also the shape of the irregularities on the material surface is important for its colonization with cells [50,137]. After grafting with the Au nanoparticles, the surface irregularities on plasma-treated PE lost their conical shape and sharpness and become rounded (figure 38), which support better spreading of cells (figure 40) and their subsequent proliferation activity.

As suggested above, not only the number of cells but also their spreading, shape and the homogeneity of their distribution on the material surface are important indicators of the cell growth activity. Although the numbers of VSMC on day 1 after seeding on pristine and modified PE were similar, these cells markedly differed in their morphology. As shown in figure 41, the cells on the pristine PE were mostly rounded, while the cells on plasma treated and especially on Au-grafted samples were spread by a significantly larger cell-material contact area, which most probably stimulated their subsequent proliferation activity. In addition, these cells were more homogeneously distributed on the material surface. On day 5 after seeding, the entire surface of the plasma-irradiated and Au-grafted PE is homogeneously covered with a confluent and dense VSMC layer, whereas the cells on the pristine PE form islands irregularly distributed on the material surface [134].

The 3T3 cells reached 100 % confluence during four days of cultivation when seeded in suggested ratio [134]. As the 3T3 cells overgrew the surface within four days, only 1 and 3 day periods were evaluated. Strikingly (and in contrast to VSMC), the pristine PE was absolutely inappropriate support material for the cultivation of 3T3 cells, as they very poorly adhered to this material and exhibited very limited proliferation during the first three days (figure 42). Moreover, granulation of the cytoplasm and apparent loss of the cell-material contact were observed after five days of cells cultivation on the pristine PE when virtually all the cells detached into the growth medium. In summary, the 3T3 cells required plasma modification for their attachment, and application of Au slightly supported both their attachment and growth (compare columns 2 and 3 in the figure 42). It can be concluded that the plasma treatment of PE enhances the adherence of both tested cell types. However, this effect is far more dramatic for 3T3 cells. On the contrary, the Au grafting exhibits a moderate positive effect on the 3T3 proliferation whilst it almost doubled the proliferation capacity of VSMC.

Chapter 8

CONCLUSION

The properties of nano-particles (crystal lattice parameter, density, magnetic and electric properties) and their chemical activity are different from those of the same material in bulk, the differences being a function of the particle size. Several experimental results and theoretical consideration, discussed above, have shown that the differences are due to the surface size- and the quantum size- effects. The first one is related to the increased fraction of the surface localized atoms in small particles, the second one with peculiar behaviour of electrons embedded in a small volume of a nano-particle.

The properties of gold layers deposited by sputtering and vacuum evaporation on PE, PET, PTFE and glass substrate were investigated and the main results were summarized. The thickness of the deposited gold layers is an increasing, nearly linear function of the deposition time for both deposition techniques, i.e. sputtering and vacuum evaporation. In sputtering onto pristine PET and PTFE the deposition rate depends on initial polymer surface morphology. While for short deposition times thicker layer is deposited onto PET, for longer deposition times the situation is reversed. Two deposition rates were observed in gold sputtering on PET, lower one for short deposition times and higher for longer ones. For vacuum evaporation on PET the deposition rate is constant for all deposition times but it is significantly lower than that for sputtering. Sheet resistance of gold layers deposited on PET and PTFE is a decreasing function of the deposition time and the layer thickness. Extremely rapid resistance decrease at a critical layer thickness, indicating transition from discontinuous to continuous layer, is observed on sputtered Au layers. For layers prepared by vacuum evaporation the sheet resistance decreases more slowly and continuity of the layer is achieved for larger layer thickness. The difference can be explained by different mechanisms of Au deposition, deposition of separated atoms by sputtering and larger atomic clusters by vacuum evaporation. For longer deposition times the resistance still decreases but much more slowly. From the measurements of the sheet resistance and reflection of electromagnetic waves it may be concluded that both sputtered and evaporated gold layers are discontinuous for thicknesses below 4 nm, continuous but heterogeneous for thicknesses from 4 to 10 nm, and continuous and homogenous for thicknesses above 10 nm. Temperature dependence of the sheet resistance, measured from 90 K to RT, strongly depends on the layer thickness. For thick layers about the resistance is an increasing function of the sample temperature, the behaviour being expected for metals. For thinner layers the sheet resistance first decreases rapidly with increasing temperature but above the temperature of about 250 K a slight resistance increase is observed. The initial decrease and the final increase of the sheet resistance with increasing temperature are typical for semiconductors and metals respectively. Extremely thin layers exhibit semiconducting properties over the whole temperature range examined.

On sputtered samples the darkening of the deposited structure and changes of colour from blue to green, both connected with increasing structures thickness were observed. After annealing the „same structures" acquire red colour. Non-zero band gap indicates semi-conducting properties of sputtered structures. Annealing leads to significant increase of the optical band gap. For sputtered gold structures the rapid decline of the sheet resistance R_s; for the annealed structures the decline appears on the structures deposited for the times above 250 s. Annealing also results in dramatic changes in the gold layer morfology and surface roughness. According to AFM images „spherolytic and hummock-like" structures are created in the gold structures by annealing.

Surface morphology of pristine polymers and polymers with deposited Au layers was characterized by AFM and SEM techniques. Differences in the surface morphology of layers with similar thickness prepared by sputtering and vacuum evaporation on different polymers are observed. Gold deposition leads to changes in the surface morphology and roughness in comparison pristine polymers, which depends on polymer initial morphology and deposition technique. For the PET substrate the layers prepared by vacuum evaporation do not change surface morphology, regardless of the layer thickness. The layers prepared by sputtering, however, change the polymer surface morphology significantly, the roughness being an increasing function of the layer thickness. Transmitting light spectra, measured on thin gold layers, exhibit absorption minimum around 500 nm which is slightly red shifted with increasing layer thickness. Pronounced absorption at longer wavelength could be attributed to surface plasmon resonance. The lattice parameter, determined by XRD technique, monotonously declines of with increasing layer thickness. The decline can be explained by the internal stress relaxation during the growth of gold clusters.

The properties of polymer substrates modified by plasma discharge or by UV laser light were investigated. Polymer surface exposed to the plasma discharge comprises low mass degradation products, free radicals and various oxidized structures, as was repeatedly proved by XPS, EPR and RBS measurements. In some cases (PE) the plasma treatment may lead to significant changes in polymer surface morphology and the surface roughness. The plasma treatment of polymers leads to a dramatic decrease of the contact angle compared to pristine polymers, which is a well-known phenomenon in the field of plasma modification. During sample aging gradual growth of the contact angle is observed, resulting of spontaneous rearrangement of degraded macromolecules and molecular fragments and other relaxation processes in the polymer surface layer.

The scratch tests show that plasma modification of PET does not affect the adhesion of subsequently deposited gold layers. The plasma modified PET surface may be grafted with dithiol. The gold layers sputtered onto plasma treated and dithiol grafted modified surface exhibit increased adhesion thanks to interaction of sputtered Au atoms with -SH group introduced by the dithiol grafting.

Irradiation by linearly polarized light from pulsed 248 nm KrF and 157 nm F_2 lasers results in the formation of coherent patterns of ripples with a lateral periodicity of the order of the wavelength of the laser light and a height of several tens of nanometer on the polymer surface. The relief was coated by a sputtered gold layer. Gold sputtering on polymer irradiated with KrF laser does not result in formation a continuous metal layer but separated "nano-wires" at the ridges of the nano-patterns are created.

Grafting of plasma modified PE with gold nano-particles results in additional wettability increase and changes in morphology and roughness. Plasma treatment and the presence of Au nano-particles were found to have a positive effect on spreading, proliferation activity and homogeneity of the distribution of VSMC cells on the PE surface. Even more pronounced positive effect of the PE plasma treatment on the adhesion of 3T3 fibroblasts is observed although following cell proliferation is stimulated only moderately. Nevertheless, the modification of PE with plasma discharge and by grafting with Au nano-particles significantly increases the attractiveness of this polymer for colonization with VSMC and 3T3 fibroblasts.

ACKNOWLEDGMENTS

The work was supported by the GA CR under the projects 106/09/0125, 108/10/1106 and 106/09/P046 and Ministry of Education of the CR under Research program No. LC 06041 and GAAS CR under the projects KAN400480701, KAN200100801 and IAA 400720710. The authors thank Ms. O.Kesselová for the technical assistance.

REFERENCES

[1]   Kaune, G; Ruderer, M.A; Metwalli, E; Wang, W; Rohlsberger, R; Roth, SV; Muller-Buschbaum, P. *Appl. Mater. Interf.*, 2009, 1, 353.

[2]   *Metallized Plastics: Fundamentals and Applications*; Mittal, KL. Ed; Marcel Dekker: New York, 1998.

[3]   *Polymers for Electronic and Photonic Applications*; Wong, CP. Ed; Academic Press: Boston, 1993.

[4]   Stutzmann, N; Tervoort, TA; Bastiaansen, K; Smith, P. *Nature*, 2000, 407, 613.

[5]   Efimenko, K; Rybka, V; Švorčík, V; Hnatowicz, V. *Appl. Phys.*, 1999, A68, 479.

[6]   Wei, B; Wang, J; Li, C; Shimada, A; Ichikawa, M; Taniguchi, Y; Kamikawa, T. *Org. Electronics*, 2008, 9, 323.

[7]   Lee, JH; Kim, SH; Kim, GH; Lim, SC; Lee, H; Jang, J; Zyung, T. *Synth. Met.* 2003, *139*, 445.

[8]   Daeil, K. *Appl. Surf. Sci.*, 2003, 218, 71.

[9]   Kim, SH; Asay, DB; Dugger, MT. *Nano Today*, 2007, 2, 22.

[10]   Houšková, J; Hoke, J; Dobiáš, J. *Czech J. Food Sci.* 1998, 16, 143.

[11]   Prosyčevas, I; Puišo, J; Tamulevičius, S; Juraitis, A; Andrulevičius, M; Čyžiūté, B. *Thin Solid Films* 2006, 495, 118.

[12]   Akamatsu, K; Tsuboi, N; Hatakenaha, Y; Deki, S. *J. Phys. Chem. B* 2000, 104, 10168.

[13]   Wilson, SA; Jourdain, RPJ; Zhang, Q. *Mater. Sci. Eng. R* 2007, 56, 1.

[14]   Choy, KL. *Progr. Mater. Sci.* 2003, 48, 57.

[15]   Zaporojtschenko, V; Zekonyte, J; Biswas, A; Faupel, F. *Surf. Sci.*, 2003, 532-535, 300.

[16] Švorčík, V; Slepička, P; Švorčíková, J; Špírková, M; Zehentner, J; Hnatowicz, V. *J. Appl. Polym. Sci.*, 2006, 99, 1698.

[17] Okay, C; Rameev, BZ; Khaibullin, RI; Okutan, M; Yildiz, F; Popok, VN; Aktas, B. *Phys. Stat. Sol.*, *A* 2006, 203, 1525.

[18] Park, P; Sung C; Yoon, SS; Nam, JD. *Thin Solid Films*, 2008, 516, 3028.

[19] Zaporojtchenko, V; Zekonyte, J; Faupel, F. *Nucl. Instrum. Meth.*, *B* 2007, 265, 139.

[20] Lia, WT; Chartersb, RB; Luther-Daviesa, B; Marb, L. *Appl. Surf. Sci.* 2004, 233, 227.

[21] Strunskus, T; Kiene, M; Willecke, R; Thran, A; von Bechtolsheim, C; Faupel, F. *Mater. Corros.* 1998, 49, 180.

[22] Metwalli, E; Couet, S; Schlage, K; Ruderer, M; Wang, W; Kaune, G; Roth, SV; Muller-Buschbaum, P. *Langmuir*, 2008, 23, 4265.

[23] Silvain, J.F.; Veyrat, A; Ehrhardt, JJ. *Thin Solid Films*, 1992, 221, 114.

[24] Seki, K; Hayashi, N; Oji, H; Ito, E; Ouchi, Y; Ishii, H. *Thin Solid Films*, 2001, 393, 298.

[25] Faupel, F. *Phys. Stat. Sol.*, *A* 1992, 134, 9.

[26] Vieth, RW. In Book *Diffusion in and through polymers*; Hanser: Munich, 1991, 38.

[27] Heuchel, M; Fritsch, D; Budd, PM; McKeown, NB; Hofmann, D.*J. Membrane Sci.*, 2008, 318, 84.

[28] Georgieva, DG; Bairda, RJ; Newazb, G; Aunera, G; Wittec, R; Herfurth, H. *Appl. Surf. Sci.*, 2004, 236, 71.

[29] Nathawat, R; Kumar, A; Kulshrestha, V; Singh, M; Ganesan, V; Phase, DM; Vijay, YK. *Appl. Surf. Sci.*, 2007, 253, 5985.

[30] Zaporojtchenko, V; Zekonyte, J; Wille, S; Schuermann, U; Faupel, F. *Nucl. Instrum. Meth.*, *B* 2005, 236, 95.

[31] Tjong, SC. *Mat. Sci. Eng. R*, 2006, 53, 73.

[32] Švorčík, V; Rybka, V; Maryška, M; Špírková, M; Hnatowicz, V. *Eur. Polym. J.*, 2004, 40, 211.

[33] Slepička, P; Rebollar, E; Heitz, J; Švorčík, V. *Appl. Surf. Sci.*, 2008, 254, 3585.

[34] Gotoh, Y; Igarashi, R; Ohkoshi, Y; Nagura, M; Akamatsu, K; Deki, S. *J. Mater. Chem.*, 2000, 10, 2548.

[35] *Handbook of Nanostructured Materials and Nanotechnology*; Nalwa, HS. Ed; Academic Press: San Diego, 2000.

[36] Bernardini, S; Masson, P; Houssa, M. *Microelectron. Eng.*, 2004, 72, 90.

[37] Goddard, JM; Hotchkiss, JH. *Prog. Polym. Sci.*, 2007, 32, 698.

[38] Ma, Z; Mao, Z; Gao, C. *Coll. Surf.*, 2007, B60, 137.

[39] Price, RL; Ellison, K; Haberstroh, KM; Webster, TJ. *J. Biomed. Mater. Res.*, 2004, A70, 129.

[40] Webster, TJ; Smith, TA. *J. Biomed. Mater. Res.*, 2005, A74, 677.

[41] Tzoneva, R; Heuchel, M; Altankov, G; Albrecht, W; Paul, D. *J. Biomater. Sci. Polym. Ed.*, 2002, 13, 1033.

[42] Bačáková, L; Filová, E; Kubies, D; Machová, L; Prokš, V; Malinová, V; Lisá, V; Rypáček, F. *J. Mater. Sci. Mater. Med.*, 2007, 18, 1317.

[43] Švorčík, V; Kolářová, K; Slepička, P; Macková, A; Hnatowicz, V. *Polym. Degr. Stab.*, 2006, 91, 1219.

[44] Mikulíková, R; Moritz, S; Gumpenberger, T; Olbrich, M; Romanin, C; Bačáková, L; Švorčík, V; Heitz, J., *Biomaterials*, 2005, 26, 5572.

[45] Heitz, J; Švorčík, V; Bačáková, L; Ročková, K; Ratajová, E; Gumpenberger, T; Bauerle, D; Dvořánková, B; Kahr, H; Graz, I; Romanin, C. *J. Biomed. Mater. Res.*, 2003, 67A, 130.

[46] Bačáková, L; Bottone, MG; Pellicciari, C; Lisá, V; Švorčík, V. *J. Biomed. Mater. Res.*, 2000, 49, 369.

[47] Gannon, CJ; Patra, CR; Bhattacharya, R; Mukherjee, P; Curley, S. *J. Nanobiotechnol,*. 2008, 6, 2.

[48] Nagesha, D; Laevsky, GS; Lampton, P; Banyal, R; Warner, C; DiMarzio, C; Sridhar, S. *Int. J. Mahomed.*, 2007, 2, 813.

[49] McGhee, AM; Procter, DJ. *Top Curr. Chem.*, 2006, 264, 93.

[50] Bačáková, L; Švorčík, V. In Book *Cell Growth Processes New Research*; Columbus, F. Ed; Nova Sci. Publish.: New York, 2008; 5-56.

[51] Base, T; Bastl, Z; Plzák, Z; Plešek, J; Carr, MJ; Malina, V; Boháček, J; Kříž, O. *Langmuir*, 2005, 21, 7776.

[52] Lovell, S; Rollinson, E. *Nature*, 1968, 218, 1179.

[53] Roduner, E. *Chem. Soc. Rev.*, 2006, 35, 583.

[54] Roduner, E. In Book *Nanoscopic Materials. Size-dependent Phenomena*; RSC Publishing: Cambridge, UK, 2006.

[55] Hornyak, GL; Dutta, J; Tibbals, HF; Rao, AK. In Book *Introduction to Nanoscience*; CRC Press: Boca Raton-London-New York, 2008.

[56] Sun, CQ. *Prog. Solid State Chem.*, 2007, 35, 1.

[57] Wang, N; Rokhlin, SI; Farson, DF. *Nanotechnology*, 2008, 19, 1.

[58] Koga, K; Ikeshoji, T; Sugawara, K. *Phys. Rew. Lett.*, 2004, 92, 115507.

[59] Rao, CNR; Kulkarni, GU; Thomas, PJ; Edwards, PP. *Chem. Eur. J.*, 2002, 8, 28.

[60] Lai, SL; Guo, JZ; Petrova, V; Ramanath, G; Allen, LH. *Phys. Rew. Lett.*, 1996, 77, 99.

[61] Soliard, C; Flueli, M. *Surf. Sci.*, 1985, 156, 487.

[62] Weissmüller, J. *J. Phys. Chem.*, *B* 2002, 106, 889.
[63] Qin, W; Nagase, T; Umakoshi, Z; Szpunar, JA. *J. Phys.-Condens. Matter* 2007, 19, 236217.
[64] Vanithakumari, SC; Nanda, KK. *Phys. Lett. A*, 2008, 372, 6930.
[65] Oberdörster, G; Oberdörster, E; Oberdörster, J. *Environ. Heealth Perspect.*, 2005, 113, 823.
[66] Salata OV. *J. Nanobiotechnol.*, 2004, 2, 3.
[67] Freitas, RA. Jr. In Book *Nanomedicine, Basic Capabilities*; Landes Bioscience: Georgetown TX, 1999.
[68] Minchin, R. *Nat. Nanotechnol.*, 2008, 3, 12.
[69] Loo, C; Lin, A; Hirsch, L; Lee, MH; Barton, J; Halas, N; West, J; Drezek, R. *Technol. Cancer. Res. Treat.*, 2004, 3, 33.
[70] Shi, X; Wang, S; Meshinchi, S; Van Antwerp, ME; Bi, X; Lee, I; Baker, JR. Jr. *Small*, 2007, 3, 1245.
[71] Siegel, J; Lyutakov, O; Kolská, Z; Šutta, P; Rybka, V; Švorčík, V. *Thin Solid Films*, submitted; [71a] Švorčík, V; Kvítek, O.; Siegel, J; Lyutakov, O; Kolská, *Appl. Phys. A*, submitted.
[72] Fischer, W; Geiger, H; Rudolf, P; Wissmann, P. *Appl. Phys.*, 1977, 13, 245.
[73] Vook, RW; Ouyang, S; Otooni, MA. *Surf. Sci.*, 1972, 29, 277.
[74] Häupl, K; Lang, M; Wissmann, P. *Surf. Interface Anal.*, 1986, 9, 27.
[75] Häupl, K; Wissmann, P; Fresenius Z. *Anal. Chem.*, 1983, 314, 337.
[76] Klabunde, KJ. In Book *Nanoscale Materials in Chemistry*; Wiley: New York, USA, 2001.
[77] Hazra, D; Datta, S; Mondal, M; Ghatak, J; Satyam PV; Gupta, AK. *J. Appl. Phys.*, 2008, 103535.
[78] Hvolbaek, B; Janssens, TVW; Clausen, BS; Falsig, H; Christensen, CH; Norskov, JK. *Nano Today*, 2007, 2, 14.
[79] Slepička, P; Kolská, Z; Náhlík, J; Hnatowicz, V; Švorčík, V. *Surf. Interf. Anal.*, 2009, 41, 741.
[80] Švorčík, V; Zehentner, J; Rybka, V; Slepička, P; Hnatowicz, V. *Appl. Phys. A*, 2002, 75, 541.
[81] Hodgman, CD. In In Book *Handbook of Chemistry and Physics*; Chemical Rubber: Cleveland, 1975, 2354.
[82] Chopra, KL. In Book *Thin Film Phenomena*; J. Wiley: New York, 1969.
[83] Mott, N. In Book *Metal insulator transitions*; Taylor & Francis: London, 1990.
[84] Švorčík, V; Endršt, R; Rybka, V; Arenholz, E; Hnatowitz, V; Černý, F. *Eur. Polym. J.*, 1995, 31, 189.

[85] Brust, M; Bethell, D; Kiely, Ch. J; Schiffrin, D. J. *Langmuir*, 1998, 14, 5425.

[86] Hunderi, O. *Surf. Sci.*, 1980, 96, 1.

[87] Kalyuzhny, G; Vaskevich, A; Schneeweiss, M; Rubinstein, I. *Chem. Eur. J.*, 2002, 8, 3850.

[88] Doremus, RH. *J. App. Phys.*, 1966, 37, 2775.

[89] Mie, G. *Ann. Phys.*, 1908, 330, 377.

[90] Bellamy, DJ; Clarke, PH. *Nature*, 1968, 218, 1179.

[91] Walton, DJ; *Chem. Phys.*, 1962, 37, 2182.

[92] Thran, A; Kiene, M; Zaporojtchenko, V; Faupel, F. *Phys. Rev. Lett.*, 1999, 82, 1903.

[93] Švorčík, V; Rybka, V; Maryška, M; Špírková, M; Hnatowicz, V. *Eur. Polym. J.*, 2004, 40, 211.

[94] Švorčík, V; Siegel, J; Slepička, P; Kotál, V; Špirková, M. *Surf. Interf. Anal.*, 2007, 39, 79.

[95] Švorčík, V; Kotál, V; Slepička, P; Blahová, O; Špirková, M; Sajdl, P; Hnatowicz, V. *Nucl. Instrum. Meth. B*, 2006, 244, 365.

[96] Lai, J; Sunderland, B; Xue, J; Yan, S; Zhao, W; Folkard, M; Michael, BD; Wang, Y. *Appl. Surf. Sci.*, 2006, 252, 3375.

[97] Gerenser, LJ. *J. Adhes. Sci. Technol.*, 1993, 7, 1019.

[98] Heitz, J; Arenholz, E; Bauerle, D. *Appl. Phys. A*, 1993, 56, 329.

[99] Mayer, G; Blanchemain, N; Dupas-Bruzek, C; Miri, V; Traisnel, M; Gengembre, L; Derozier, D; Hildebrand, HF. *Biomaterials*, 2006, 27, 553.

[100] Pandiyaraj, KN; Selvarajan, V; Deshmukh, RR; Gao, Ch. *Vacuum*, 2009, 83, 332.

[101] Ardelean, H; Petit, S; Laurens, P; Marcus, P; Arefi-Khonsari, F. *Appl. Surf. Sci.*, 2005, 243, 304.

[102] Zhu, XL; Liu, SB; Man, BY; Xie, CQ; Chen, DP; Wang, DQ; Ye, TC; Liu, M. *Appl. Surf. Sci.*, 2007, 253, 3122.

[103] Tsakalakos, L. *Mater. Sci. Eng. R*, 2008, 62, 175.

[104] Lazare, S; Benet, P. *J. Appl. Phys.*, 1993, 74, 4953.

[105] Arenholz, E; Heitz, J; Himmelbauer, M; Bäuerle, D. *Proc. SPIE*, 1996, 2777, 90.

[106] Csete, M; Eberle, R; Pietralla, M; Marti, O; Bor, Z. *Appl. Surf. Sci.*, 2003, 208-209, 474.

[107] Birnbaum, M. *J. Appl. Phys.*, 1965, 36, 3688.

[108] Geil, P.H. *Europhys. Conf. Abstr.*, 1988, 12D, 22.

[109] Bäuerle, D. In Book *Laser Processing and Chemistry*; 3rd ed; Springer: Berlin- Heidelberg-New York, 2000.

[110] Sipe, JE; Young, JS; Preston, Van Driel, HM. *Phys. Rev. B*, 1983, 27, 1141.

[111] Csete, M; Bor, Z. *Appl. Surf. Sci.*, 1998, 133, 5.

[112] Bolle, M; Lazare, S. *J. Appl. Phys.*, 1993, 73, 3516.

[113] Siegel, J; Slepička, P; Heitz, J; Kolská, Z; Sajdl, P; Švorčík, V. *Appl. Surf. Sci.*, 2010, 256, 2205.

[114] Kotál, V; Švorčík, V; Slepička, P; Bláhová, O; Šutta, P; Hnatowicz, V. *Plasma Proc. Polym.*, 2007, 4, 69.

[115] Kim, KS; Ryu, CM; Park, CS; Sur, GS; Park, CE. *Polymer*, 2003, 44, 6287.

[116] Wu, G; Paz, MD; Chiussi, S; Serra, J; Gonzalez, P; Wang, YJ; Leon, B. *J. Mater. Sci. Mater. Med.*, 2009, 20, 597.

[117] Guimond, S; Wertheimer, MR. *J. Appl. Polym. Sci.*, 2004, 94, 1291.

[118] Siegel, J; Řezníčková, A; Chaloupka, A; Slepička, P; Švorčík, V. *Rad. Eff. Def. Sol.* 2008, 163, 779.

[119] Wilson, DJ; Williams, RL; Pond, RC. *Surf. Interface. Anal.*, 2001, 31, 385.

[120] Tahara, M; Cuong, NK; Nakashima,Y. *Surf. Coat. Tech.*, 2003, 174, 826.

[121] Choi, Y; Kim, J; Paek, K; Ju, W; Hwang, YS. *Surf. Coat. Tech.*, 2005, 193, 319.

[122] Kihlman Øiseth, S; Krozer, A; Kasemo, B; Lausmaa, J. *Appl. Surf. Sci.*, 2002, 202, 92.

[123] Švorčík, V; Kotál, V; Siegel, J; Sajdl, P; Macková, A; Hnatowicz, V. *Polym. Degr. Stab.*, 2007, 92, 1645.

[124] Švorčík, V; Hnatowicz, V; Stopka, P; Bačáková, L; Heitz, J; Ryssel. H. *Rad. Phys. Chem.*, 2001, 60, 89.

[125] Švorčík, V; Kotál, V; Bláhová, O; Špírková, M; Sajdl, P; Hnatowicz, V. *Nucl. Instrum. Meth. B*, 2006, 244, 365.

[126] Barr, TL. In Book *Modern ESCA;* CRC Press: Boca Raton FL, 1994.

[127] Bertrand, P; Lambert, P; Travaly, Y. *Nucl. Instrum. Meth. B*, 1997, 131, 71.

[128] Le, QT; Pireaux, JJ; Caudano, R; Leclere, P; Lazzaroni, R. *J. Adhes. Sci. Technol.*, 1998, 12, 999.

[129] Švorčík, V; Kotál, V; Slepička, P; Bláhová, O. Šutta, P; Hnatowicz, V. *Polym. Eng. Sci.*, 2006, 46, 1326.

[130] Švorčík, V; Chaloupka, A; Záruba, K; Král, V; Bláhová, O; Macková, A. *Nucl. Instrum. Meth. B*, 2009. 267, 2484.

[131] Miller, DC; Thapa, A; Haberstroh, KM; Webster, TJ. *Biomaterials*, 2004, 25, 53.

[132] Webster, TJ; Ergun, C; Doremus, RH; Bizios, R. *J. Biomed. Mater. Res.*, 2000, 51, 475.

[133] Curry, AC; Crow, M; Wax, A. *J. Biomed. Opt.*, 2008, 13, 14022.

[134] Švorčík, V; Kasálková, N; Slepička, P; Záruba, K; Bačáková, L; Pařízek, M; Ruml, T; Macková, A. *Nucl. Instrum. Meth. B*, 2009, 267, 1904.

[135] Chu, PK; Chen, JY; Wang, LP; Huang, N. *Mater. Sci. Eng. R*, 2002, 36, 143.

[136] Ročková, K; Švorčík, V; Bačáková, L; Dvořánková, B; Heitz, J. *Nucl. Instrum. Meth. B*, 2004, 225, 275.

[137] Bačáková, L; Walachová, K; Švorcik, V; Hnatowitz, V. *J. Biomater. Sci. Polym. Ed.*, 2001, 12, 817.

[138] Bačáková, L; Filová, E; Rypáček, F; Švorčík, V; Starý, V. *Physiol. Res.*, 2004, 53, S35.

INDEX